T0280561

CAMBRIDGE LIBRARY COLLECTION

Books of enduring scholarly value

Physical Sciences

From ancient times, humans have tried to understand the workings of the world around them. The roots of modern physical science go back to the very earliest mechanical devices such as levers and rollers, the mixing of paints and dyes, and the importance of the heavenly bodies in early religious observance and navigation. The physical sciences as we know them today began to emerge as independent academic subjects during the early modern period, in the work of Newton and other 'natural philosophers', and numerous sub-disciplines developed during the centuries that followed. This part of the Cambridge Library Collection is devoted to landmark publications in this area which will be of interest to historians of science concerned with individual scientists, particular discoveries, and advances in scientific method, or with the establishment and development of scientific institutions around the world.

Photographs of Stars, Star-Clusters and Nebulae

A geologist and fellow of the Royal Astronomical Society, Isaac Roberts (1829–1904) made significant contributions to the photography of star-clusters and nebulae. By championing reflecting rather than refracting telescopes, Roberts was able to perceive previously unnoticed star-clusters, and was the first person to identify the spiral shape of the Great Andromeda Nebula. Roberts' use of a telescope for photographing stars, and a long exposure time, provided greater definition of stellar phenomena than previously used hand-drawings. Although Roberts' conclusions about the nature of the nebulae he photographed were not always correct, the book is significant for the possibilities it suggests for nebular photography. Published in 1893 and 1899, the two-volume *Photographs of Stars* represents the summation of his work with his assistant W.S. Franks at his observatory in Crowborough, Sussex. Volume 2 contains 29 plates of stars, and his conclusions about their origins and nature.

Photographs of Stars, Star-Clusters and Nebulae

Together with Records of Results Obtained in the Pursuit of Celestial Photography

VOLUME 2

ISAAC ROBERTS

CAMBRIDGE
UNIVERSITY PRESS

CAMBRIDGE UNIVERSITY PRESS

Cambridge, New York, Melbourne, Madrid, Cape Town, Singapore,
São Paolo, Delhi, Dubai, Tokyo

Published in the United States of America by Cambridge University Press, New York

www.cambridge.org
Information on this title: www.cambridge.org/9781108015233

This edition first published 1899
This digitally printed version 2010

ISBN 978-1-108-01523-3 Paperback

PHOTOGRAPHS

OF

STARS, STAR-CLUSTERS AND NEBULÆ,

TOGETHER WITH

RECORDS OF RESULTS OBTAINED IN THE PURSUIT OF CELESTIAL PHOTOGRAPHY.

BY

ISAAC ROBERTS, D.Sc., F.R.S.,

Honorary Member of the American Philosophical Society;

Fellow of the Royal Astronomical and of the Geological Societies of London, etc.

VOLUME II.

London:

"KNOWLEDGE" OFFICE, 326, HIGH HOLBORN, W.C.

THE COLLOTYPE PLATES BY THE LONDON STEREOSCOPIC COMPANY.

PREFACE.

MY intention, in the pages following, is to convey in brief, and I hope clear form, my views concerning some of the results already obtained by the aid of photography in the elucidation of celestial problems, the complete solution of which cannot for many years yet be obtained; and I may here quote from the preface to the volume, issued in the year 1893, of *A Selection of Photographs of Stars, Star-Clusters, and Nebulæ*, the following paragraphs, which are applicable also to the present volume.

"It has been my aim, in publishing the photographs and descriptive matter introduced in the following pages, to place data in the hands of astronomers, for the study of astronomical phenomena, which have been obtained by the aid of mechanical, manipulative, and chemical processes of the highest order at present attainable; and that such data should be, as regards the photographs, free from all personal errors."

"The photographs portray portions of the Starry Heavens in a form at all times available for study, and identically as they appear to an observer aided by a powerful telescope and clear sky for observing."

In the processes employed for obtaining the photographic illustrations contained in this volume the same instruments have been used, and the same care has been exercised in the production of the illustrations of the various objects as in the first volume; but owing to improvements in the manufacture of photographic films, and to the extended data now available beyond that which had been obtained up to the year 1893, when the first volume was published, certain deductions concerning the evolution of stellar systems are now permissible which six years ago would have been justly considered premature.

The evidence now published appears to me of so striking a character that it should no longer be withheld from discussion. In presenting it I have endeavoured to avoid personal predilections or bias of any kind, and would only ask for the same fairness in any criticism to which my views may be submitted.

ISAAC ROBERTS.

STARFIELD, CROWBOROUGH HILL,
SUSSEX.
December, 1899.

CONTENTS.

LIST OF THE PLATES.

INSTRUMENTS.

The instruments employed were the silver-on-glass reflector of 20-inches aperture and 98-inches focus; and a specially-made triplet portrait-lens of 5-inches aperture and 19·22 inches focus, by Messrs. Cooke & Sons.

LIST OF ABBREVIATIONS ADOPTED IN THIS WORK

N.G.C.—*A New General Catalogue of Nebulæ and Clusters of Stars*, by Dr. J. L. E. Dreyer. Published in the *Memoirs of the Royal Astronomical Society*, Vol. XLIX., Part I.

IND. C.—*Index Catalogue of Nebulæ* found in the years 1888 to 1894, with Notes and Corrections to the *New General Catalogue*, by Dr. J. L. E. Dreyer. Published in the *Memoirs of the Royal Astronomical Society*, 1895, Vol. LI., Part VII.

G.C.—Sir J. F. W. Herschel's *Catalogue of Nebulæ and Clusters of Stars*. Published in the *Philosophical Transactions* of the Royal Society for the year 1864, Vol. CLIV., Part I.

h.—Sir J. F. W. Herschel's *Observations of Nebulæ and Clusters of Stars*. Published in the *Philosophical Transactions* of the Royal Society for the year 1833.

I.R. PHOTOS.—*A Selection of Photographs of Stars, Star Clusters and Nebulæ*, by Isaac Roberts, D.Sc., F.R.S. Published in 1893.

D.M.—*Durchmusterung*, by Argelander. Bonn, 1859.

n. N.—North. *s.* S.—South. *p.*—Preceding. *f.*—Following.

F.S.—Fiducial Star, marked (·), (··), (∴), (∷).

Kn.—*Knowledge*. Monthly Serial. Published by Messrs. Witherby & Co., London.

M.N.—*Monthly Notices* of the Royal Astronomical Society.

THE NEGATIVES.

The negatives, from which the photographs taken with the 20-inch reflector have been enlarged, measure 10 centimètres square, and one Equatorial Degree upon them measures 44·2 millimètres. The photographic field of the 5-inch lens is 15 degrees in diameter.

All the plates have been enlarged by photographic methods from the negatives to the scales given in the letterpress referring to each subject.

EPOCH OF THE FIDUCIAL STARS—1900.

Certain stars on the plates are marked with dots, numbering from one dot to four or five, thus (::), and the co-ordinates of these stars are given for the epoch 1900. The co-ordinates are to be accepted as approximately accurate, since the positions of some of the stars have not been determined with the highest degree of precision. They will, however, suffice for finding the stars in space, and for determining their relative positions on the plates and also on the Celestial Sphere.

TABLE.

For converting the measured Right Ascensions of the stars shown on those photographs which are enlarged to the scale of 1 millimetre to 30 seconds of arc into intervals of time at each Degree in Declination between the Equator and the Pole.

Declination.	1 Millimetre =	Declination.	1 Millimetre =	Declination.	1 Millimetre =
	s.		s.		s.
0°	2·00 in R.A.	30°	2·30 in R.A.	60°	3·98 in R.A.
1	2·00 ,,	31	2·32 ,,	61	4·12 ,,
2	2·00 ,,	32	2·35 ,,	62	4·26 ,,
3	2·00 ,,	33	2·37 ,,	63	4·41 ,,
4	2·00 ,,	34	2·40 ,,	64	4·56 ,,
5	2·00 ,,	35	2·44 ,,	65	4·74 ,,
6	2·00 ,,	36	2·47 ,,	66	4·92 ,,
7	2·01 ,,	37	2·50 ,,	67	5·10 ,,
8	2·01 ,,	38	2·53 ,,	68	5·32 ,,
9	2·02 ,,	39	2·56 ,,	69	5·57 ,,
10	2·02 ,,	40	2·60 ,,	70	5·84 ,,
11	2·04 ,,	41	2·64 ,,	71	6·12 ,,
12	2·04 ,,	42	2·68 ,,	72	6·45 ,,
13	2·05 ,,	43	2·72 ,,	73	6·82 ,,
14	2·05 ,,	44	2·77 ,,	74	7·29 ,,
15	2·06 ,,	45	2·82 ,,	75	7·77 ,,
16	2·07 ,,	46	2·87 ,,	76	8·26 ,,
17	2·09 ,,	47	2·92 ,,	77	8·76 ,,
18	2·10 ,,	48	2·98 ,,	78	9·50 ,,
19	2·11 ,,	49	3·04 ,,	79	10·42 ,,
20	2·12 ,,	50	3·10 ,,	80	11·59 ,,
21	2·14 ,,	51	3·17 ,,	81	12·81 ,,
22	2·15 ,,	52	3·24 ,,	82	14·31 ,,
23	2·17 ,,	53	3·31 ,,	83	16·59 ,,
24	2·18 ,,	54	3·39 ,,	84	19·21 ,,
25	2·20 ,,	55	3·47 ,,	85	22·65 ,,
26	2·22 ,,	56	3·56 ,,	86	29·20 ,,
27	2·24 ,,	57	3·66 ,,	87	38·42 ,,
28	2·26 ,,	58	3·76 ,,	88	60·82 ,,
29	2·28 ,,	59	3·87 ,,	89	121·66 ,,

DETERIORATION OF THE NEGATIVES.

If a reason had to be given in addition to the obvious advantage of this method of publication by printing the photographs with permanent ink it would be afforded by the fact that the records obtained by photography are peculiarly liable to be lost by accidental breakage of the glass negatives. Besides this there is the certainty that after the lapse of a limited number of years the gelatine films will become discoloured; the images will fade, and the faint stars and the faint nebulosities will entirely disappear from view.

I have had within my own experience proofs that the faint stars fade from the films, and will give the following examples:—On the 15th February, 1886, a photograph was taken of the region of the sky with the co-ordinates R.A. 9h. 40m. Dec. North 72°·0 at the centre of the plate; exposure 15m.; area of the plate four square degrees.

Shortly after the photograph was taken I counted 403 star-images on the negative; and on 29th May, 1895, I again counted the stars on the same negative, and found only 272. Therefore, stars to the number of 131 had entirely disappeared from the film in the course of nine and a-quarter years.

Another photograph of identically the same region was taken with an exposure of fifteen minutes on the 22nd March, 1886, and soon after that date I counted 364 stars upon the negative. In May, 1895, I again counted the stars, and found only 234. Therefore, 130 star-images had disappeared from the film in the interval of nine and one-fifth years. These are only two of several instances I could adduce to prove that faint star-images fade from the negatives.

It follows from this evidence that the following photographs, which are printed in permanent form with printer's ink, though they fail to show all the faint stars and faint nebulosities that are visible on the original negatives, yet the objects are depicted to the degree of faintness that would be represented by stars of about the 17th magnitude; and it will be conceded that this is a great advance upon the records made from eye observations and drawings by hand-work, which were the only methods of astronomical recording till recent years, when the photographic method was introduced.

After the lapse of a few years, when other photographs of regions of the sky which are coincident with those here charted have been taken, the work of correlating may be profitably undertaken, for there will be ample material available, in a reliable form, for the astronomical measurers, computers, and deducers of laws. We in these days can only desire that we might live to see the results of their labours.

EFFECTS OF ATMOSPHERIC GLARE AND OF DIFFRACTION UPON THE FILMS OF PHOTOGRAPHIC PLATES.

Very sensitive gelatine films, such as are used in photography, when exposed during several hours to the sky in taking stellar photographs, become more or less darkened during development. The darkening is chiefly due to atmospheric glare caused by star-light; and the nebulous circles seen round the bright stars are caused by the glare and by diffraction effects produced by the objectives, or mirrors, of the instruments employed in photographing.

I have made some experiments to enable us to judge to what extent the glare and diffraction affect the finished photographs, a summary of which experiments may be given here. They were made by exposing simultaneously plates in the 20-inch reflector, the 5-inch lens camera, and to the sky in a blackened box, measuring 7-inches square by 12-inches in height, with the open end exposed to the zenith, the exposures respectively being made during precisely equal intervals of time. The plates were selected so as to be equal in sensitiveness, and the development was performed in a similar manner in each coincident trial.

The plates exposed in the box were 6-inches square, and equal areas on each of them were (1) left uncovered; (2) covered with black paper; (3) covered with different thicknesses of polished plate glass. The plates when developed showed the comparative effects of the unobstructed full sky glare as well as the effects of the application of complete and partial covering with plates of glass or with sensitometer figured scales.

The following are some of the results obtained:—The plates referred to as measuring 6-inches square and exposed in the box to the sky at the zenith showed, after development, those parts which were covered with black paper as if they were clear glass—no photographic effect being perceptible—whilst the parts covered with polished plate glass of the respective thickness of 11·26, 22·52, and 45 millimetres, showed gradations in the darkening of the films proportional to the absorption by the glass and reflections from the surfaces. The parts which were exposed uncovered showed the films to be darkened by the sky glare to the density of the images of stars of about the 16th magnitude on plates exposed during the same time for two and a-half hours in the 20-inch reflector. Similar results were obtained by exposing plates during two and a-quarter hours in the 20-inch reflector, and simultaneously to the sky under a figured sensitometer scale.

These experiments point to a source of spurious nebulosity, and also to the probable limit of the applicability of the photographic method in the delineation of faint stars and faint nebulosity. So far as I am able at present to judge, under the atmospheric conditions prevalent in this country, the limit of the photographic method of delineation will be reached at stellar, or nebular, light of the feebleness of about 18th magnitude stars. The reason for this inference is that the general illumination of the atmosphere by star-light concentrated upon the film by the instrument will mask the light of objects that are fainter than about 18th magnitude stars.

Of course, the inferences here stated cannot be considered as the final solution of the problem, for photographs will have to be taken by various instruments under clear atmospheric conditions and the results checked by eye observations, aided by the most powerful telescopes, before finality can properly be pronounced.

ARRANGEMENT OF THE PLATES.

The plates are arranged in classes or groups so as to indicate apparent physical relationship between them, and the Right Ascensions are, as far as practicable, given in the order of time within each group.

The edge next to the printed heading on each plate is the *south*, and the lower edge the *north* ; the right is the *following*, and the left the *preceding* edge.

The scales of the photographs, which are given in the letterpress, are such that by eye alignments of the stars, without the application of measuring instruments, changes which have taken place in their positions or in the structures of the nebulosities, if these changes should not be less than about five seconds of arc in extent, could be detected by comparing corresponding dual plates in this simple manner. The examination and comparison of stars, both as regards their positions and magnitudes, could thus be made in a single day though they should number several thousands on the dual photographs.

Besides this alignment method, measurements by scale and compasses, or by a réseau on glass or other transparent substance, or by a rectangular L-shaped metal rule divided into millimètres on both limbs, or by the superposition of the plates upon each other, are obvious methods available for detecting changes in the position angles and magnitudes of the stars shown on the photographs.

METHOD FOR MICRO-PUNCTURING THE PHOTOGRAPHIC DISCS OF STARS ON PLATES TO OBTAIN THEIR OPTICAL CENTRES FOR PURPOSES OF MEASUREMENT.

On photographs which have been taken with long exposures the stellar images are large, so that a considerable amount of uncertainty is involved in bisecting them with accuracy when measuring position angles of and distances between such stars ; and the method of making several measurements and resetting the instrument for each one in order to obtain a mean position for the centre is troublesome. But if with a fine needle point the centre of the photographic disc could be accurately punctured, one careful measurement made from such a well-defined centre point would be more reliable than the mean of several approximate measurements.

The following method of puncturing the star discs I have found in practice to give satisfactory results :—

A microscope is taken that has a revolving stage and also a sub-stage for holding and centering an achromatic condenser. By removing the front lens of the condenser and replacing it with a brass cap in which is inserted a fine needle point at its centre, we obtain a fixed point which can by the screw adjustments of the condenser be accurately placed in the optical axis of the microscope. In the ocular is placed, through the ordinary micrometer slit, a finely ruled glass réseau or a spider-line cross with the intersection of the cross, or of any two lines of the réseau at right angles brought into the optical centre of the instrument. The stellar negative is then placed, film downwards, upon the revolving stage and weighted to prevent slipping; the star-image is then (under sufficient magnifying power) brought to the centre of the field of the ocular, which will be at the intersection of the réseau lines or other fixed point. We have thus a combination of three fixed visible points in line in the optical axis of the microscope, and all we have further to do is to bring up the needle point, by the carrier of the condenser focussing screw, until it impinges upon the star-image on the film ; by which operation a microscopic puncture is made in the film from which accurate measurements can always be made. These operations are simple and the results satisfactory.

A convenient modification of this method is to puncture only the large star-images and the irregular outlined condensations of nebulosity involved in the convolutions and central nuclei of spiral nebulæ, and then using a double micrometer eye-piece, with two double spider lines moving at right angles and at any distance from each other, across the field of the ocular as described in the *Monthly Notices* of the R.A.S., Vol. 49, Pl. I., pp. 6-7, the centre of the star-images will coincide with the centre of the square formed by the four spider lines when they are placed tangentially to the image.

STAR CATALOGUES AND PHOTOGRAPHIC CHARTS.

The catalogues of stars are numerous, as will be seen by referring to the list given in Chambers' "Handbook of Descriptive Astronomy," where 170 are enumerated between that by Hipparchus in the year B.C. 128 and the year A.D. 1876; besides these there are catalogues of nebulæ, and atlases or charts of stars.

In considering these records one is impelled to ask the questions—Have these vast stores of computative and descriptive literature, the product of great energy expended in physical and mechanical operations, and of much thought, been utilized in the advancement of astronomical knowledge to a degree commensurate with the labour and cost of their production?

Or, have the astronomers been deterred from undertaking the work of correlation, on a comprehensive scale, because they know that there is a considerable margin of probable error in all these records—particularly in those of earlier date than the middle of the present century—they therefore judge it would be unprofitable to devote their time to making the necessary examinations. They know that when they find differences to exist in the records, it would be uncertain whether they were objective, or were only the result of human errors.

In my own experience I have found differences to exist between photographs and carefully prepared modern charts of stars—differences in the position angles, in distances, in the magnitudes of the stars and in the structure and extent of nebulosities, which were most probably due to errors in charting.

With photographic charts, on the contrary, when the time interval between any two photographs of the same region of the sky has extended to twenty or thirty years and upwards, all the uncertainties here suggested will vanish; and when a difference is found to exist it will be known to be objective and real, and can with confidence be dealt with mathematically.

It might appear to some that the correlation of the stars in the catalogues respectively would be as easy and reliable as the examination on the photographs; but a little consideration will show this would not be the case; for, in the catalogues errors may have occurred in observing, in recording, in computing, and in printing, each of which would be absent on the photographs.

The labour of comparing the stars singly in the catalogues, and in deducing results from these comparisons would be enormously greater than that of the comparison of photographs; especially if the method of super-position of the negatives, or of copies of them made to the same scale, were adopted. In this way hundreds of stars could be examined as regards position, distance, and magnitude, in less time than tens could be examined by the aid of catalogues, and the results would be reliable.

The range of the examination is also largely in favour of the photographs; for stars to the faintness of 18th magnitude, and as far apart as two degrees can be examined upon them; whereas, there are few catalogues containing stars fainter than the 9th or 10th magnitudes.

DURATION OF THE EFFECTIVE EXPOSURES GIVEN TO PHOTOGRAPHIC PLATES.

It is a general opinion that the longer the time a sensitive film is exposed, in a photographic instrument, under clear atmospheric conditions, the greater will be the number of stars and the extent of nebulosity imprinted upon the film. But so far as my experience enables me to judge, after twelve years' use of the 20-inch reflector, and more than two years' use of an excellent and specially-made portrait lens combination of 5-inches aperture and 19-inches focus, the limit of photographic effect is reached sometime within ten to twelve hours on clear nights, and with very sensitive films, in the 20-inch reflector. With the 5-inch lens very much longer exposures may be given before the darkening of the films, by atmospheric glare and diffraction effects, reach the same degree of density as in the reflector.

The photographic effect produced by the 5-inch lens with an exposure of two or three hours and upwards is about two stellar magnitudes less than that given by the reflector in the same time and with films of equal sensitiveness. It would, therefore, appear that, given sufficient time, the atmospheric glare would, in both instruments, mask or extinguish the light of faint stars and faint nebulosity, which is provisionally assumed to be equal to that of 18th magnitude stars. When that limit has been reached no fainter light-effect than this would be imprinted on the films; and upon these premises the questions in the following section require consideration.

ARE THE MILLIONS OF STARS AND THE NUMEROUS NEBULOSITIES, WHICH ARE NOW KNOWN TO EXIST, LIMITED IN NUMBER AND EXTENT; AND DO THEY CONSEQUENTLY INDICATE THAT THE UNIVERSE OF WHICH THE SOLAR SYSTEM CONSTITUTES A PART IS ONLY ONE MEMBER OF A GREATER STELLAR UNIVERSE?

The questions involved in the heading have engaged the attention of astronomers in the past, but does the photographic method contribute evidence more reliable in character than that formerly available?

Let us consider the evidence which has up to the present time been obtained by photography, remembering that the last dozen years covers the whole interval during which it has been accumulated:—

(A) Eleven years ago photographs of the Great Nebula in *Andromeda* were taken with the 20-inch reflector and exposures of the plates during intervals up to four hours; and upon some of them were depicted stars to the faintness of 17th to 18th magnitude, and nebulosity to an equal degree of faintness. The films of the plates obtainable in those days were less sensitive than those that have been available during the past five years, and during this period photographs of the nebula with exposures up to four hours have been taken with the 20-inch reflector. No extensions of the nebulosity, however, nor increase in the number of the stars, can be seen on the later rapid plates than were depicted upon the earlier slower ones, though the star-images and the nebulosity have greater density on the later plates.

(B) The Great Nebula in *Orion* has been photographed with the 20-inch reflector at frequent intervals between the years 1886 and 1898 with exposures varying between one minute of time and seven hours thirty-five minutes; yet the stars are not more numerous or the extensions of the nebulosity greater on the latter than are shown on a plate of like sensitiveness which had been exposed during ninety minutes only; the difference exhibited was that of density.

(C) The group of the *Pleiades* has been photographed with the 20-inch reflector during numerous intervals between 1886 and 1898 with exposures of between one minute and twelve hours. The results are that only the same faint stars and nebulosity seen upon plates which have had an exposure of one-and-a-half hours are depicted upon those which have been exposed during ten or twelve hours.

(D) Several photographs of the region of the *Milky Way* in *Cygnus* (Plate 5) have been taken with the 20-inch reflector between the years 1886 and 1898, and on comparing two of them (one with an exposure of 60 minutes and the other with 2h. 35m.) no fainter stars could be found on one than on the other; there is no reason for suspecting that this result was due to some abnormal condition in either case, and it has been confirmed by photographs which have been taken of other areas in the sky.

Here then is evidence founded upon photographs of objects at different altitudes and positions in the sky, all obtained under favourable conditions with an instrument of considerable power and on films of a high degree of sensitiveness, which I think may be accepted as demonstrations of the accuracy of the surmises of astronomers in the past that the part of the Starry Universe visible from the earth is limited in extent, and that notwithstanding the enormous assistance afforded by the photographic method we are again brought to a check because of the inadequacy of the powers we possess to enable us to peer beyond that part of space in the midst of which we are placed; and though we know that it extends over countless millions of miles we seem to be no nearer than our predecessors were in descrying a boundary. Space appears to us to be infinite and beyond the grasp of our mental capacity.

Whilst drawing these inferences we must endeavour to realise in however small a degree the bewildering extent of that part of the Stellar Universe which is within our range when we avail ourselves of the highest optical, chemical and other powers that we can call to our aid.

THE EVOLUTION OF STELLAR SYSTEMS.

A century has elapsed since Laplace suggested that the sun, and planets, might have been evolved out of nebulous matter, but his imagination did not lead him to realize the much larger idea that stellar systems might also have been evolved from matter similar in its constitution.

I now propose to submit a series of photographic copies of my original negatives from which we may obtain strong evidence, if not complete demonstration, of the evolution of stellar systems.

The first part of the series will consist of photographs of rich fields of stars, and of clusters showing various degrees of concentration; these will be followed by a series of the spiral nebulæ; some of them symmetrical in form, and others less symmetrical though clearly spiral. Following these again will be a number of nebulæ of circular, annular, and irregular forms, and, lastly, nebulæ consisting of large areas of cloud-like matter having irregular structural characteristics.

The appearances to which I now wish to draw special attention in the examination of these photographs are the numerous curves and lines of stars that are associated together in separate groups. The stars are of nearly equal magnitude; of approximately equal distances apart in each group, and the groups are independent of each other and of the surrounding stars.

These appearances are so numerous and persistent amongst the stars that their attribution to chance coincidences cannot be entertained, and we are irresistibly driven to accept as the cause that the stars in each group respectively are closely related to one another and have been formed of similar material at a relatively co-equal epoch of time.

By similar material I mean material in a state of dissociation without regard to its chemical constitution; for such is the feebleness of the light received from these stars and nebulæ that the spectroscope cannot at present be employed in the determination of the particular elementary substances that generate their light, and it will probably long remain impracticable to show, either visually or photographically, the spectra of light so feeble as that of stars of 16th to 18th magnitude.

INFERENCES SUGGESTED BY EXAMINATION OF THE PHOTOGRAPHS.

The next step to be taken is to verify the accuracy of the above statements regarding the grouping of the stars into lines and curves; assuming that the examiner has, like myself, been convinced of their reality. As a test of this, as well as an example let us examine Plates 2 to 9, upon which the eye readily detects many groups of stars arranged in lines and in curves, each of them containing several stars; similar configurations to these can be seen on the other plates, and if I had chosen to print hundreds of others that are in my cabinets, each covering four square degrees in the sky, similar configurations would be seen upon them.

This persistency of the lines and curves of stars on the plates leaves no room for doubt that they are the effects of physical causes, and cannot be due to coincidence only; and when the photographs of the spiral and other nebulæ are examined a reasonable explanation of the formation of the curves and lines will be made manifest.

It is not my intention to submit elaborate arguments, or mathematical formulæ, in the discussion of the photographic evidence contained in this and in the first volume of my photographs—these will in the future, when a sufficient interval of time has elapsed, occupy the thoughts of the correlators, the measurers, the computers and of the mathematicians—my aim is now to point out the evidence, and the relationships to each other of the several classes of objects that are found depicted, untouched by hand-work, upon the photographs.

Assuming then that we have seen the various curves and lines of stars, which are undeniably visible upon the photographic charts, we shall further find that they are strikingly similar in appearance to those condensations of nebulous or cometic matter which are involved in the convolutions of the spiral nebulæ depicted on Plates Nos. 10 to 18 following.

There are also to be seen stars, apparently in a complete state of development, scattered over the surfaces of the most prominent of the nebulæ, but it will be observed that they do not conform with the trends of the spirals nor with the curves of the nebulous stars involved in them. This fact I apprehend to be strong evidence that they are independent of the nebulæ—that they are not in any way involved in the nebulosity, but are seen by us either in front or else in space beyond the nebulæ. If they were beyond them their light would have to penetrate through the nebulosity, and we should therefore expect it to be duller in character and the margins of the stars be surrounded by more or less dense nebulous rings; but these effects are not traceable in the photo-images, and we are consequently led to adopt the alternative inference that they are between us and the nebulæ. If they were involved in the nebulosity they would conform with the trends of the convolutions and appear like nebulous stars.

It must be borne in mind when comparing the photographs of the spiral nebulæ with the star charts that the scale of the former is two and a-half times that of the latter.

The idea will occur to many that the spectroscope should be an available instrument, in skilful hands, to throw much light upon the question of the progressive development of nebulosity into stars. But when we realise that the stellar condensations and the nebulosity produce light of such feeble character that it may be estimated to be only equal to that of stars of the 16th or 17th magnitude, it will be evident that no form of spectroscope has hitherto been devised that can show the spectra of these stars or even of those of the faintness of 12th magnitude. Therefore there is no probability of assistance being obtainable by aid of the spectroscope.

We shall have to rely upon the observations of physical changes in the structures of the nebulosities, of movements amongst the stars, and variations in their light intensity for the information we are seeking.

Here again we are met with obstructions connected with the element of time ; for such are the distances in space of the objects, concerning which we are seeking for knowledge, that a life-time is a period too short to enable observers to detect more than microscopic changes that may have taken place amongst them. But though the recorders of the present day, like those in the past, may not themselves see and rejoice in the fruits of their labours, yet they will have the satisfaction of handing down to their successors photographic records that will be free from all personal errors and be available for investigations within that unit—the *Galactic System*—which is dimly within the grasp of their imagination in the vast unknown and inconceivable expanse of space.

Let us now examine the photograph of one of the largest and brightest of the spiral nebulæ, *Messier* 51 *Canum Venaticorum* (Plate 15) together with the diagram (Plate 16) and the tabulated measurements of the position angles and distances of the stars and star-like condensations on pages 25 and 109.

I would first point out that to Lord Rosse is due the credit of recognising the spiral character of this and many other nebulæ. In the year 1872-74 he measured, with all possible care, the position angles and distances from the nuclei of some of the stars and star-like condensations involved in this nebula; but unfortunately the measurements did not extend to stars so distant from the nebula as to be free from suspicion of their physical connection with its system and therefore uninfluenced by its movements. We must not, however, forget the great, perhaps insurmountable, difficulties observers in those days had to contend with in measuring with the necessary accuracy such long distances. Had this been practicable in the year 1872 we should now be in a position to deduce some very interesting conclusions concerning this nebula.

Notwithstanding the absence of these essential records to enable us to arrive at exact conclusions, we are not left without indications that changes have taken place in this nebula during the interval of twenty-six years (1872 to 1898) ; for on comparing the measurements of the star positions given by Lord Rosse in his *Obs of Neb. and Cl. of Stars*, p. 131, with those that can be recognised to correspond with them on the photograph Pl. 15, and more clearly shown on the diagram Pl. 16, we find the differences to be as shown in the following table, and they are probably due to actual changes in the nebula, and amongst the stars, during forty-seven years (1851 to 1898).

Lord Rosse states (reference above) : "From his measurements made in the years 1872-74, Dr. Copeland has calculated the following *final results.* The position angles have been reduced to 1851 (Corr. + 4′·0) and are very reliable, the distances are poor." My thanks are due to Dr. Copeland for reducing the position angles of the stars to the epoch 1898 for comparison with the measurements I have made from the photograph taken in that year. The results are shown in the following table.

TABLE

Showing probable changes that have taken place in the nebula between the years 1851 and 1898.

	Lord Rosse's Measurements. Epoch 1898.		Measurements on the Negative. Epoch 1898.		Differences + = Roberts in Excess of Rosse.	
	Pos. Angle.	Dist. from N.	Pos. Angle.	Dist. from N.	Pos. Angle.	Distance.
N n	15° 2′	260″·3	16° 52′	261″	+ 1° 50′	+ 0″·7
2	46 20	281 ·5	50 4	290	+ 3 44	+ 8 ·5
4	108 30	241 ·0	109 32	241	+ 1 2	0 ·0
5	162 5	97 ·2	163 59	100	+ 1 54	+ 2 ·8
6	188 52	245 ·5	190 52	242	+ 2 0	− 3 ·5
7	210 28	170 ·0	211 38	170	+ 1 10	0 ·0
10	218 41	125 ·7	220 20	132	+ 1 39	+ 6 ·3
8	222 52	203 ·9	224 39	204	+ 1 47	+ 0 ·1
17	228 1	114 ·1	228 50	116	+ 0 49	+ 1 ·9
9	230 8	84 ·0	231 12	85	+ 1 4	+ 1 ·0
13	275 51	229 ·7	277 29	226	+ 1 38	− 3 ·7
11	277 39	112 ·0	277 29	111	− 0 10	− 1 ·0
14	304 41	236 ·9	307 5	233	+ 2 24	− 3 ·9
15	308 4	182 ·3	310 7	181	+ 2 3	− 1 ·3

It would appear from the examination of this table that only two of the fourteen stars measured by Lord Rosse are clearly involved in the substance of the nebula—these are the nucleus (n) of the northern head and the condensation marked 10; or 56 on my list (p. 109).

Five of the stars numbered by Lord Rosse 2, 4, 6, 11, and 15 (they are numbered on my list 15, 33, 48, 68, and 75) are outside the convolutions, and may therefore not be physically connected with the nebula.

The remaining seven stars are in the spaces between the convolutions, and we must at present consider it doubtful if they are physically connected with it.

By this analysis we would infer that the whole nebula has revolved on its nucleus as a centre to the extent of one hundred and four minutes of arc during the interval of forty-seven years, whilst the stars (which are not certainly involved in the nebula, but probably within the sphere of its influence) have moved in the same direction to the extent of from sixty-two minutes to (in one instance) two hundred and twenty-four minutes in the forty-seven years. One of the stars (No. 11) has apparently retrograded to the extent of ten minutes of arc.

We must, for the present, accept this evidence with due caution; for the differences, or some portion of them, may be due to unavoidable errors arising from the extreme difficulty of measuring with great accuracy (with the instruments available fifty years ago) the position angles, and distances of these very faint objects with irregular outlines. Even on the photographs there is some uncertainty in finding the centres for measurement purposes of the irregular nebulous condensations involved in the convolutions.

There are spiral nebulæ, in the convolutions of which the star-like condensations present various degrees in the symmetry of disc and density of image, as will be observed on several of the photographs, and the general aspect of the nebulæ themselves shows various degrees in symmetrical development.

If it is admitted that the evidence now produced proves, or even strongly indicates, the evolution of stars from the material of which spiral nebulæ are composed, from what source is the material derived? And is it gaseous or meteoritic? The existence in space of these different forms of matter is abundantly proved by the photographs of vast luminous clouds, and it is corroborated by observations of meteoric matter and of cosmical dust falling continuously upon the earth.

We also know that no object whatever has been found in the whole range of space that does not partake of a motion of translation, and in some instances such motions probably exceed two hundred miles in each second of time. We also know of the existence of dark bodies, probably extinct suns and planets, but of their frequency and number we can form no conception.

The movements of these bodies are also in various directions with reference to each other. Hence it is not an unreasonable assumption that collisions must take place amongst them—that stars may collide with stars, or with nebulæ, or that nebulæ may collide with each other.

What would be the effect of such collisions? Let us suppose that a sun collided centrally with another of equal mass, and both were moving with known celestial velocity from opposite directions in space. (A) The effect of the collision would be to produce a discoidal mass of gaseous and comminuted material spreading out in circular form at right angles with the line of impact, while a mass of matter at the centre would remain to form a nucleus to a newly-formed spiral nebula, which would be composed of the wreckage of the two suns. The axial rotation of each sun at collision, and the angle at which they met, would also affect the motions and distribution of the matter in the resulting discoidal mass.

(B) If a sun rushed into a nebula from a direction perpendicular with its plane the result would be a temporary brightening of one hemisphere of the sun, and a vortical disturbance in the nebula in and around the point of impact.

(C) If two nebulæ collided with each other the result would be an increase of mass, and of brightness, with a very complicated structure.

These postulates refer only to the simplest forms of collisions, but they are not outside the range of probability or possibility if we may judge by the evidence furnished by the photographs, and corroborated by original negatives in my cabinets, which can only at present be referred to there.

There are numerous historical records of the more or less sudden appearance of new stars, and even in our own time similar phenomena have occurred. On the strength of this collective evidence we cannot dismiss as incredible the inferences that I have briefly submitted in this chapter, which are, that collisions between bodies in space may be one of the causes leading to the evolution of new stellar systems out of the wreckage of those pre-existent.

We now proceed with the description and examination of the photographs; always bearing in mind that the original negatives show the evidence more clearly than is possible to be reproduced on paper prints.

Plate 1.

CLUSTERS H VII. 46 & VI. 31 CASSIOPEIÆ.

CLUSTER H VII. 61 PERSEI.

PLATE I.

Clusters H VII. 46 and VI. 31 Cassiopeiæ.

H VII. 46. R.A. 1h. 37m. 10s. Dec. N. 61° 23′·0.

H VI. 31. R.A. 1h. 39m. 11s. Dec. N. 60° 44′·5.

The photograph covers the region between R.A. 1h. 34m. 56s. and R.A. 1h. 41m. 57s. Declination between 60° 15′·3 and 61° 37′·8 North.

Scale—1 millimetre to 30 seconds of arc.

Co-ordinates of the Fiducial Stars marked with dots for the epoch A.D. 1900.

Star (.)	D.M. No. 308—	Zone + 60°	...	R.A. 1h. 35m. 17·9s.	...	Dec. N. 60° 31′·6	...	Mag. 6·5
„ (··)	„ „ 312	„ 60°	...	„ 1h. 36m. 10·5s.	...	„ 60° 54′·9	...	„ 6·3
„ (∵)	„ „ 327	„ 61°	...	„ 1h. 41m. 9·5s.	...	„ 61° 21′·2	...	„ 8·7

The photograph was taken with the 20-inch reflector on January 15th, 1893, between sidereal time 2h. 34m. and 3h. 34m., with an exposure of the plate during sixty minutes.

REFERENCES.

H VII. 46 N.G.C. 654. G.C. 387. *h*. 145.

H VI. 31 „ 663. G.C. 392.

Sir J. Herschel, in the G.C., describes H VII. 46 as a cluster; irregular figure; rich; 1 star 6·7, stars 11-14. H VI. 31 is described as a cluster; bright; large; extremely rich; stars pretty large.

Lord Rosse (*Obs. of Neb. and Cl.*, p. 21) describes H VII. 46 as a coarse cluster, a good many stars 8th mag. in it, and some down to 13th mag.

The photograph shows stars down to about 16th mag.; several of them are apparently double and triple. There is no nebulosity shown in either of the clusters, and there are several areas shown on the plate without any stars in them so bright as 16th magnitude.

PLATE I.

Cluster H VII. 61 Persei.

R.A. 4h. 7m. 38s. ; Dec. N. 50° 59′·2.

The photograph covers the region between R.A. 4h. 4m. 47s. and R.A. 4h. 10m. 54s. Declination between 50° 17′·8 and 51° 40′·3 North.

Scale—1 millimètre to 30 seconds of arc.

Co-ordinates of the Fiducial stars marked with dots for the epoch A.D. 1900.

Star ( )	D.M.	No. 926	—Zone+50°	...	R.A. 4h. 5m. 5·4s.	...	Dec. N.	50° 54′·4	...	Mag.	8·0
,, (··)	,,	,, 941	,, 50°	...	,, 4h. 7m. 31·0s.	...	,,	50° 25′·8	...	,,	7·4
,, (·:)	,,	,, 963	,, 50°	...	,, 4h. 9m. 40·3s.	...	,,	50° 37′·1	...	,,	7·0

The photograph was taken with the 20-inch reflector on January 15th, 1893, between sidereal time 4h. 30m. and 5h. 30m., with an exposure of the plate during sixty minutes.

REFERENCES.

N.G.C. 1528. G.C. 820.

Sir J. Herschel, in the G.C., describes the cluster as bright; very rich ; considerably compressed.

Lord Rosse (*Obs. of Neb. and Cl.*, p. 40) describes it as rich ; little compressed ; the stars have a tendency to a spiral arrangement. 30′ *s.f.* is a very red star 9½ mag. ; stars 10—11 magnitude.

The photograph shows the cluster to consist of widely scattered stars, and Lord Rosse's suggestion of a spiral arrangement amongst them is indicated; several of them have apparent comites. There is no indication of nebulosity in the group, and the stars are shown down to about the 16th magnitude. There are also vacant places without any stars so bright as about 16th magnitude, both in the cluster and in the surrounding area.

Plate 2.

CLUSTER H VI. 5 ORIONIS.

CLUSTER M. 46 & NEBULA H IV. 39 ARGÛS.

PLATE II.

Cluster H VI. 5 Orionis.

R.A. 6h. 8m. 10s. Dec. N. 12° 50′·1.

The photograph covers the region between R.A. 6h. 6m. 6s. and R.A. 6h. 10m. 5s. Declination between 12° 10′·8 and 13° 33′·3 North.

Scale—1 millimètre to 30 seconds of arc.

Co-ordinates of the Fiducial stars marked with dots for the epoch A.D. 1900.

Star (.)	D.M. No. 1164	—Zone+13°	...	R.A. 6h. 7m. 14·5s.	...	Dec. N. 13° 9′·8	...	Mag. 9·1
„ (··)	„ „ 1074	„ 12°	...	„ 6h. 8m. 59·6s.	...	„ 12° 31′·3	...	„ 8·6
„ (∵)	„ „ 1174	„ 13°	...	„ 6h. 9m. 30·4s.	...	„ 13° 4′·0	...	„ 9·2

The photograph was taken with the 20-inch reflector on February 23rd, 1897, between sidereal time 6h. 16m. and 7h. 16m., with an exposure of the plate during sixty minutes.

REFERENCES.

N.G.C. 2194. G.C. 1383.

Sir J. Herschel, in the G.C., describes the cluster as large; rich; gradually very much compressed in the middle.

Lord Rosse (*Obs. of Neb. and Cl.*, p. 52) found a loose cluster and a more concentrated one in it; also two others not quite so remarkable, about 16′ and 21′ *following*, both crescent-shaped, convexity *preceding*.

The photograph shows the stars down to about 16th magnitude, with configurations both within and around the cluster strongly suggestive of its origin from a spiral nebula but without any apparent nebulosity remaining in it. The two other clusters referred to by Lord Rosse are also suggestive of a spiral origin, and stars with apparent *comites* are numerous on this area.

PLATE II.

Cluster M. 46 and Nebula H IV. 39 Argûs.

R.A. 7h. 37m. 14s. Dec. S. 14° 35′·2.

The photograph covers the region between R.A. 7h. 35m. 11s. and R.A. 7h. 39m. 10s. Declination between 13° 55′·3 and 15° 17′·8 South.

Scale—1 millimètre to 30 seconds of arc.

Co-ordinates of the Fiducial Stars marked with dots for the epoch A.D. 1900.

Star	D.M. No.	Zone	R.A.	Dec. S.	Mag.
Star (.)	D.M. No. 2082	—Zone—14° ...	R.A. 7h. 35m. 48·7s. ...	Dec. S. 15° 1′·9 ...	Mag. 5·4
,, (··)	,, ,, 2099	,, 14° ...	,, 7h. 36m. 28·5s. ...	,, 14° 49′·9 ...	,, 8·7
,, (∵)	,, ,, 2158	,, 14° ...	,, 7h. 38m. 3·6s. ...	,, 14° 7′·2 ...	,, 8·6
,, (::)	,, ,, 2171	,, 14° ...	,, 7h. 38m. 20·3s. ...	,, 14° 46′·4 ...	,, 8·8

The photograph was taken with the 20-inch reflector on February 24th, 1894, between sidereal time 6h. 2m. and 7h. 32m., with an exposure of the plate during ninety minutes.

REFERENCES.

N.G.C. 2437. G.C. 1564. *h* 463.

Sir J. Herschel, in the G.C., describes the cluster as a remarkable object; very bright; very large; very rich; with a planetary nebula involved.

The photograph shows the cluster as well as the surrounding regions to be densely covered with stars down to about 17th magnitude. They are shown in lines and curves of a remarkable character. Many have apparent *comites*; many are double and multiple; but pages might be filled with descriptive matter, which, at best, would convey only an imperfect reference to that which is shown on the plate itself, especially if a magnifier is used in its examination; but notwithstanding the number of stars that can be seen on the plate the faintest ones shown on the negative are lost in the reproduction.

The nebula is also referred to in the descriptive matter accompanying Plate 18.

CLUSTER H VI. 37 ARGÛS.

CLUSTER M. 24 CLYPEI.

PLATE III.

Cluster H VI. 37 Argûs.

R.A. 7h. 55m. 9s. Dec. S. 10° 20′·8.

The photograph covers the region between R.A. 7h. 53m. 19s. and R.A. 7h. 57m. 10s. Declination between 9° 49′·7, and 11° 12′·2 South.

Scale—1 millimètre to 30 seconds of arc.

Co-ordinates of the Fiducial Stars marked with dots for the epoch A.D. 1900.

Star								
Star (.)	D.M. No. 2313—Zone — 10°	...	R.A. 7h. 54m. 3·9s.	...	Dec. S. 10° 40′·4	...	Mag. 8·7
„ (··)	„ „ 2324 „ 9°	...	„ 7h. 55m. 33·5s.	...	„ 10° 3′·8	...	„ 8·2
„ (∴)	„ „ 2332 „ 10°	...	„ 7h. 55m. 54·1s.	...	„ 10° 58′·5	...	„ 9·0

The photograph was taken with the 20-inch reflector on February 27th, 1894, between sidereal time 6h. 24m. and 7h. 54m. with an exposure of the plate during ninety minutes.

REFERENCES.

N.G.C. 2506. G.C. 1611. *h* 480.

Sir J. Herschel, in the G.C., describes the cluster as pretty large; very rich; compressed; stars 11—20 magnitude.

Lord Rosse (*Obs. of Neb. and Cl.*, p. 63) records the results of nine observations made between the years 1849 and 1856. He saw a spiral appearance about the brightest star, and some unresolved milkiness about the central star, very curious black spaces, smaller stars so confused together as to give a nebulous look to parts of the cluster.

The photograph shows the cluster to be a remarkable aggregation of faint stars in the midst of crowded surroundings of similar stars, which cover the sky over an area of several degrees. The black spaces referred to by Lord Rosse are very conspicuous, especially on the negative, and resemble those in the globular cluster *Messier* 13 *Herculis.* There are also branching lines and curves of stars very similar to those seen in spiral nebulæ, but there is no indication of nebulosity.

The clusters *Herschel VI.* 33, 34 *Persei* and *Messier* 11 *Antinoi* resemble that here referred to, and they also are free from nebulosity.

The resemblances here pointed out strongly suggest the existence of a relationship between spiral nebulæ and globular clusters, for, if the former are due to the collision of bodies in space, the latter would represent the condensation into stars of the central nucleus resulting from the collision; and the curves, lines, and branches represent the condensation into stars of the remainder of the gigantic wreckage, widely spreading in a discoidal form.

PLATE III.

Cluster M. 24 Clypei.

R.A. 18h. 12m. 35s. Dec. S. 18° 27′·4.

The photograph covers the region between R.A. 18h. 10m. 36s. and R.A. 18h. 14m. 34s. Declination between 17° 45′·0 and 19° 7′·5 South.

Scale—1 millimètre to 30 seconds of arc.

Co-ordinates of the Fiducial Stars marked with dots for the epoch A.D. 1900.

Star (.) D.M.	No. 4886—Zone—18° ...	R.A. 18h. 11m. 37·1s. ...	Dec. S. 18° 30′·0 ...	Mag. 6·7
„ (··) „	„ 4896 „ 18° ...	„ 18h. 12m. 51·0s. ...	„ 18° 39′·6 ...	„ 7·3
„ (∵) „	„ 4900 „ 18° ...	„ 18h. 12m. 56·4s. ...	„ 18° 12′·5 ...	„ 8·2

The photograph was taken with the 20-inch reflector on August 14th, 1895, between sidereal time 18h. 57m. and 20h. 57m., with an exposure of the plate during two hours.

REFERENCES.

N.G.C. 6603 G.C. 4397. *h* 2004.

Sir J. Herschel, in the G.C., describes the cluster as remarkable; very rich; very much compressed; round; stars of 15th magnitude.

The photograph shows the cluster to be in form somewhat resembling a horse-shoe, with the open side facing the *n.p.*, and the stars, both in it and in the regions around, are in lines and curves of various forms. The stream of the *Milky Way* is well shown on the photograph with dark spaces in which are few stars and in some places none. In photographs on a small scale, such as those taken with portrait lenses, these vacancies, in which there is little or no light, form a contrast with the areas crowded with stars which produce atmospheric glare, and this fact has given rise to the erroneous idea that the stars in the greater part of the *Milky Way* are immersed in nebulosity. But photographs taken on a large scale dissipate such appearances by showing the star-images so widely separated that the effects of star-light and atmospheric glare are hardly perceptible, and cannot be misleading.

The brightest star on the plate is about 6·8 magnitude and the faintest about

17th magnitude. The short white line which is shown in the middle of the cluster consists of five or six stars so close together that their photographic images overlap ; and on several other parts of the plate, where the stars appear elongated, it is due to the overlapping of the discs of double or multiple stars.

It will be noticed that there are remarkable aggregations of stars on several parts of the plate which might be classed as clusters.

Plate 4.

CLUSTER M. 11 ANTINOI.

STARS IN REGION OF Ḫ VI. 38 AQUILÆ.

PLATE IV.

Cluster M. 11 Antinoi.

R.A. 18h. 45m. 42s. Dec. S. 6° 23′·2.

The photograph covers the region between R.A. 18h. 43m. 50s. and R.A. 18h. 47m. 41s. Declination between 5° 41′·2 and 7° 3′·7 South.

Scale—1 millimetre to 30 seconds of arc.

Co-ordinates of the Fiducial stars marked with dots for the epoch A.D. 1900.

Star		D.M. No.				R.A.		Dec. S.		Mag
Star (.)	D.M. No.	4922—Zone	−6°	...	R.A.	18h. 44m. 19·8s.	...	Dec. S. 6° 1′·3	...	Mag 6·8
„ (··)	„ „	4924 „	6°	...	„	18h. 44m. 41·4s.	..	„ 6° 36′·7	...	„ 9·1
„ (∵)	„ „	4935 „	6°	...	„	18h. 46m. 10·8s.	...	„ 6° 4′·8	...	„ 9·6
„ (::)	„ „	4936 „	6°	...	„	18h. 46m. 16·7s.	...	„ 6° 24′·7	...	„ 9·5

The photograph was taken with the 20-inch reflector on August 10th, 1896, between sidereal time 19h. 36m. and 21h. 6m., with an exposure of the plate during ninety minutes.

REFERENCES.

N.G.C. 6705. G.C. 4437. *h* 2019.

Sir J. Herschel, in the G.C., describes the cluster as a remarkable object; very bright; large; irregularly round; rich; stars 9 to 11 magnitude.

Lord Rosse (*Obs. of Neb. and Cl.*, p. 51) describes the cluster as curiously broken up into groups—one star two or three classes brighter than the rest.

The photograph shows stars down to about the 17th magnitude, and that they are crowded both in the cluster and on the southern half of the plate, but on the northern half they are much less numerous. The vacancies or regions where the stars are few and faint are very striking, and, by the contrast they make with the crowded parts, have been mistaken for extensive nebulosity on photographs taken on a small scale. In the cluster as well as on parts of the surrounding region the stars are arranged in lines and curves with lane-like spaces between them which strongly suggest the idea that the cluster was formed from a spiral nebula.

PLATE IV.

Stars in the Region of H VI. 38 Aquilæ.

R.A. 19h. 26m. 49s. Dec. N. 9° 0′ 7.

The photograph covers the region between R.A. 19h. 25m. 29s. and R.A. 19h. 29m. 20s. Declination between 8° 27′·8 and 9° 50′·3 North.

Scale—1 millimètre to 30 seconds of arc.

Co-ordinates of the Fiducial stars marked with dots for the epoch A.D. 1900.

Star (.) D.M. No. 4124—Zone+9°	...	R.A. 19h. 25m. 47·5s.	...	Dec. N. 9° 24′·6	...	Mag. 8·5
„ (··) „ „ 4139 „ 9°	...	„ 19h. 27m. 26·7s.	...	„ 9° 7′·7	...	„ 7·0
„ (∵) „ „ 4144 „ 8°	...	„ 19h. 29m. 11·8s.	...	„ 8° 57′·9	...	„ 8·3

The photograph was taken with the 20-inch reflector on August 14th, 1896, between sidereal time 18h. 59m. and 20h. 59m., with an exposure of the plate during two hours.

REFERENCES.

N.G.C. 6804. G.C. 4499. *h* 2043.

Sir J. Herschel, in the G.C., describes H VI. 38 as considerably bright ; small ; irregularly round ; resolvable.

Lord Rosse (*Obs. of Neb. and Cl.*, p. 153) says, 4 stars in nebula ; 2 more on *p.* edge ; pretty bright and of irregular outline . . . . . 3 stars plainly seen ; suspect more smaller ones.

The photograph shows the region to be crowded with stars down to 17-18 magnitude ; many apparently double and multiple stars, arranged in lines and curves, too numerous to be particularised. Many of them being very faint are lost in the reproduction. The object is not a star-cluster as recorded by Herschel, but two bright stars surrounded by nebulosity—shown on the plate 11 of arc *s.p.* star (··). The three stars referred to by Lord Rosse are shown—two of them being on the *s.p.* margin of the nebulosity ; the other is just clear of the nebulosity and has a small bright *comes*.

CHART OF STARS IN CYGNUS.

PLATE V.

Chart of Stars in Cygnus.

The photograph covers the region between R.A. 19h. 41m. 8s. and R.A. 19h. 49m. 9s. Declination between 34° 25′·4 and 36° 32′·7 North.

Scale—1 millimètre to 31·6 seconds of arc, equal to MM. Henry's scale.

Co-ordinates of the Fiducial stars marked with dots for the epoch A.D. 1900.

Star	(.)	D.M. No.	3786—Zone +	35°	...	R.A.	19h. 41m. 59·9s.	...	Dec. N.	35° 51′·0	...	Mag. 7·0
,,	(··)	,,	3701 ,,	34°	...	,,	19h. 42m. 7·4s.	...	,,	34° 45′·9	...	,, 6·6
,,	(·.·)	,,	3727 ,,	34°	...	,,	19h. 45m. 0·2s	...	,,	35° 3′·4	...	,, 7·0
,,	(::)	,,	3826 ,,	35°	...	,,	19h. 46m. 39·2s.	...	,,	35° 50′·9	...	,, 7·5
,,	(·:·)	,,	3744 ,,	36°	...	,,	19h. 48m. 37·1s.	...	,,	36° 10′·3	...	,, 6·3

The photograph was taken with the 20-inch reflector on September 10th, 1898, between sidereal time 20h. 7m. and 22h. 42m., with an exposure of the plate during two hours and thirty-five minutes.

Several photographs of this region have been taken with the 20-inch reflector between August, 1886, and September, 1898, and one of them (which was taken on August 14th, 1887, and was published in my first volume of *I.R. Photos.*, pl. 43, p. 111), forms an accurate chart of upwards of 16,000 stars, the photo-plate being exposed during sixty minutes only.

The photograph here reproduced covers the same region and area of the sky as that just referred to, and the stars upon it number upwards of 30,000. A photograph of this region was also taken on the 17th August, 1895, with an exposure of sixty minutes only, and on comparing the original negative with that of the plate annexed, all the star-images, down to the faintest, are found to be visible on both, notwithstanding the fact that one had sixty minutes' exposure, and the other two hours and thirty-five minutes. The sensitiveness of the films and the quality of the sky during both exposures may be considered equal, and the only noticeable difference in the star-images on the two negatives is greater density on that with the longer exposure.

This is an illustration (one of several which could be adduced) pointing to the probability that all the stars existent upon this area of the sky are charted upon these negatives. Of course, the faintest star-images are lost on the photo-enlargement on paper.

The inferences we may draw from these results are that this, apparently, one of the most densely crowded star areas in the *Milky Way*, can be seen through, and that nothing visible within the limit of our powers lies beyond; and further, that the limit in space of the *Galactic System* is now probably revealed to us. But it would be unscientific to dogmatise concerning this large question before confirmation is obtained by photographs taken with other instruments under the best atmospheric conditions, and until charts of the faintest stars have been made by aid of the most powerful visual telescopes, and the results rigorously compared on the charts and on negatives before final decision can be given.

Is it too much to expect that the possessors of such instruments will utilize them in the solution of the questions above referred to?

Plate 6.

CLUSTER M. 71 SAGITTÆ.

CLUSTER H VII. 59 CYGNI.

PLATE VI.

Cluster M. 71 Sagittæ.

R.A. 19h. 49m. 17s. Dec. N. 18° 31′·0

The photograph covers the region between R.A. 19h. 47m. 16s. and R.A. 19h. 51m. 17s. Declination between 17° 49′·9 and 19° 12′·4 North.

Scale—1 millimètre to 30 seconds of arc.

Co-ordinates of the Fiducial Stars marked with dots for the epoch A.D. 1900.

Star (.) D.M. No. 4276—Zone + 18° ...	R.A. 19h. 47m. 55·1s. ...	Dec. N. 18° 24′·3 ...	Mag. 6·6
„ (··) „ „ 4293 „ 18° ...	„ 19h. 49m. 59·9s. ...	„ 18° 49′·9 ...	„ 9·2
„ (∴) „ „ 4295 „ 18° ...	„ 19h. 50m. 9·0s. ...	„ 18° 24′·1 ...	„ 9·1

The photograph was taken with the 20-inch reflector on July 20th, 1898, between sidereal time 17h. 55m. and 19h. 25m., with an exposure of the plate during ninety minutes.

REFERENCES.

N.G.C. 6838. G.C. 4520. *h* 2056.

Sir J. Herschel, in the G.C., describes the cluster as very large; very rich; pretty much compressed; stars 11-16 magnitude.

The photograph shows a cluster in which the curves and arrangements of stars closely resemble those in a spiral nebula with, possibly, a very faint trace of nebulosity about the centre. The region around the cluster is densely crowded with stars, down to about 17th magnitude, arranged in remarkable curves and lines which are very suggestive of having been produced by the effects of spiral movements. Many of the stars appear to have *comites* and multiple stars connected with them.

By using a lens to examine the photograph it will be seen that the faint stars as well as the brighter ones are spirally arranged, but of course many of the faint stars visible on the negative are lost in the reproduction.

The work of years might be devoted to the study of this one photograph, though it covers less than three square degrees of the sky, and it will be readily seen that there are other aggregations of stars on the photograph that might be described as clusters.

PLATE VI.

Cluster H VII. 59 Cygni.

R.A. 20h. 0m. 30s. Dec. N. 43° 42′·9.

The photograph covers the region between R.A. 19h. 58m. 8s. and R.A. 20h. 3m. 26s. Declination between 43° 3′·7 and 44° 26′·2 North.

Scale—1 millimètre to 30 seconds of arc.

Co-ordinates of the Fiducial stars marked with dots for the epoch A.D. 1900.

Star (.)	D.M. No 3457—Zone + 43°	...	R.A. 19h. 58m. 30·0s.	... Dec. N. 43° 50′·2	...	Mag. 6·8
„ (··)	„ „ 3466 „ 43°	...	„ 20h. 0m. 2·0s.	... „ 43° 30′·7	...	„ 9·1
„ (∵)	„ „ 3483 „ 43°	...	„ 20h. 2m. 50·2s.	... „ 44° 1′·3	...	„ 9·0

The photograph was taken with the 20-inch reflector on September 12th, 1895, between sidereal time 20h. 22m. and 21h. 22m., with an exposure of the plate during sixty minutes.

REFERENCES.

N.G.C. 6866. G.C. 4544. *h* 2066.

Sir J. Herschel, in the G.C., describes the cluster as large; very rich; considerably compressed.

The photograph shows the cluster to consist of stars down to about the 16th magnitude, widely scattered, and that they are arranged in curves and in lines both in the cluster and in the surrounding regions; many of them have *comites* and many are multiple stars.

Plate 7.

STARS IN REGION OF *h* 2107 CYGNI.

STARS IN REGION OF H VI. 24 CYGNI.

PLATE VII.

Stars in the Region of *h* 2107 Cygni.

R.A. 21h. 7m. 40s. Dec. N. 45° 16′·2.

The photograph covers the region between R.A. 21h. 4m. 59s. and R.A. 21h. 10m. 25s. Declination between 44° 31′·3 and 45° 53′·8 North

Scale—1 millimètre to 30 seconds of arc.

Co-ordinates of the Fiducial stars marked with dots for the epoch A.D. 1900.

Star (.) D.M. No. 3718—Zone + 44°	...	R.A. 21h.	6m. 23·7s.	...	Dec. N. 45°	5′·7	...	Mag. 6·8
„ (··) „ „ 3436 „ 45°	...	„ 21h.	7m. 17·0s.	...	„	45° 41′·5	...	„ 8·2
„ (∵) „ „ 3456 „ 45°	...	„ 21h.	10m. 5·5s.	...	„	45° 11′·6	...	„ 7·5

The photograph was taken with the 20-inch reflector on September 15th, 1895, between sidereal time 20h. 35m. and 21h. 35m., with an exposure of the plate during sixty minutes.

REFERENCES.

N.G.C. 7039. G.C. 4645.

Sir J. Herschel, in the G.C., describes the cluster as very large; pretty rich; extended; stars 10th magnitude.

Lord Rosse (*Obs. of Neb. and Cl.*, p. 160) states that the cluster is very large; pretty rich; stars 11th magnitude; more than 15′ long in the direction *n.p.* to *s.f.*, and is about 8′ broad.

The photograph shows a rich field of stars rather than a cluster; and that the surrounding regions have arrangements of the stars in curves and lines. On the *s.p.* and *n.f.* sides there are large areas containing very few stars.

PLATE VII.

Stars in the Region of H̥ VI. 24 Cygni.

R.A. 21h. 9m. 15s. Dec. N. 42° 5′·1.

The photograph covers the area between R.A. 21h. 7m. 33s. and R.A. 21h. 11m. 39s. Declination between 42° 8′·8 and 43° 14′·8 North.

Scale—1 millimètre to 24 seconds of arc.

Co-ordinates of the Fiducial stars marked with dots for the epoch A.D. 1900.

Star (.)	D.M. No. 4037—Zone + 41°	... R.A. 21h. 8m. 37·7s.	... Dec. N. 41° 56′·8	... Mag. 9·0
„ (··)	„ „ 4044 „ 41°	... „ 21h. 9m. 56·4s.	... „ 42° 5′·1	... „ 9·1
„ (∵)	„ „ 4024 „ 42°	... „ 21h. 11m. 19·3s.	... „ 42° 32′·1	... „ 8·2

The photograph was taken with the 20-inch reflector on October 4th, 1896, between sidereal time 21h. 35m. and 23h. 5m., with an exposure of the plate during ninety minutes.

REFERENCES.

N.G.C. 7044. G.C. 4648. *h* 2110.

Sir J. Herschel, in the G.C., describes the cluster as very faint; pretty large; very rich; very compressed; stars 15th to 18th magnitude.

Lord Rosse (*Obs. of Neb. and Cl.*, p. 160) describes the cluster as exceedingly faint, with five or six arms branching off. In finder eyepiece it is as faint as *h* 2084, and has a nebulous look.

The photograph shows the cluster to be generally as described by Lord Rosse, but there is no appearance of nebulosity. The branching arms, consisting of curves and lines of very faint stars, are suggestive of a spiral origin; curves and lines of stars also appear in the surrounding regions together with several spaces on the photograph with few if any stars in them.

Plate 8.

STARS IN REGION OF Ħ VII. 52 CYGNI.

CLUSTER Ħ VI. 32 CYGNI.

PLATE VIII.

Stars in the Region of H VII. 52 Cygni.

R.A. 21h. 25m. 51s. Dec. N. 46° 39′·4.

The photograph covers the region between R.A. 21h. 26m. 37s. and R.A. 21h. 32m. 8s. Declination between 47° 19′·5 and 48° 42′·0 North.

Scale—1 millimètre to 30 seconds of arc.

Co-ordinates of the Fiducial stars marked with dots for the epoch A.D. 1900.

Star								
Star (.)	D.M. No. 3352—Zone + 46°	...	R.A. 21h. 27m. 47·4s.	...	Dec. N. 46° 49′·4	...	Mag. 8·5
„ (··)	„ „ 3467 „ 47°	...	„ 21h. 29m. 32·8s.	...	„ 47° 20′·7	...	„ 8·4
„ (∴)	„ „ 3375 „ 46°	...	„ 21h. 31m. 15·6s.	...	„ 46° 38′·5	...	„ 7·8

The photograph was taken with the 20-inch reflector on September 18th, 1895, between sidereal time 20h. 29m. and 21h. 59m., with an exposure of the plate during ninety minutes.

REFERENCES.

N.G.C. 7082. G.C. 4673. *h* 2122.

Sir J. Herschel, in the G.C., describes the cluster as large; considerably rich; little compressed; stars 10 to 13 magnitude.

Lord Rosse (*Obs. of Neb. and Cl.*, p. 161) says that it is a large space thickly studded with stars, but no definite cluster.

The photograph confirms Lord Rosse's observation that there is no definite cluster, and further shows that the whole region, covering several degrees in area, is so densely studded with stars arranged in lines and curves, with *comites* and multiple stars apparently connected with them, that it has the appearance of being one vast cluster.

Many of the curves and lines of stars are remarkable in form and are separated by spaces with few or no stars in them.

PLATE VIII.

Cluster H VI. 32 Cygni.

R.A. 21h. 27m. 5s. Dec. N. 51° 8′·6.

The photograph covers the region between R.A. 21h. 24m. 9s. and 21h. 30m. 15s. Declination between 50° 29′·1 and 51° 51′·6 North.

Scale—1 millimètre to 30 seconds of arc.

Co-ordinates of the Fiducial Stars marked with dots for the epoch A.D. 1900.

Star (.) D.M. No. 3072—Zone + 51°	...	R.A. 21h. 25m. 42·8s.	... Dec. N. 51° 39′·6	...	Mag. 8·2
„ (··) „ „ 3355 „ 50°	...	„ 21h. 27m. 41·0s.	... „ 50° 48′·6	...	„ 8·7
„ (∵) „ „ 3082 „ 51°	...	„ 21h. 28m. 40·9s.	... „ 51° 26′·7	...	„ 7·5

The photograph was taken with the 20-inch reflector on September 21st, 1895, between sidereal time 20h. 49m. and 21h. 49m., with an exposure of the plate during sixty minutes.

REFERENCES.

N.G.C. 7086. G.C. 4676. *h* 2124.

Sir J. Herschel, in the G.C., describes the cluster as considerably large; very rich; pretty compressed; stars 11-16 magnitude.

Lord Rosse (*Obs. of Neb. and Cl.*, p. 161) describes it as an oval or circle of stars, with two curved arcs or tails. Three bright stars, 10-11 mag.

The photograph shows the cluster to be in a region where the surrounding stars are not very numerous. It consists of lines and curves of stars, which are very suggestive of a spiral structure, and they are shown down to about the 16th mag.; many multiple stars, and stars with *comites*, are visible in the cluster and in the surrounding region. The spaces where the stars are very few in number, or altogether absent, are conspicuous on the plate, and still more strikingly so on the original negative.

Plate 9.

NEBULA H IV. 75 & CLUSTER H VII. 66 CEPHEI.

CLUSTER H VI. 30 CASSIOPEIÆ.

PLATE IX.

Cluster H VII. 66 and Nebula H IV. 75 Cephei.

R.A. (Neb.) 21h. 40m. 41s. Dec. N. 65° 38′·7.
„ (Cl.) 21h. 43m. 32s. „ 65° 20′·3.

The photograph covers the region between R.A. 21h. 37m. 10s. and R.A. 21h. 46m. 21s. Declination between 64° 48′·5 and 66° 11′·0 North.

Scale—1 millimètre to 30 seconds of arc.

Co-ordinates of the Fiducial stars marked with dots for the epoch A.D. 1900.

Star (.) D.M. No. 1626—Zone + 65°	...	R.A. 21h. 37m. 34s.	...	Dec. N. 65° 17′·2	... Mag. 8·5
„ (··) „ „ 1634 „ 65°	...	„ 21h. 40m. 12s.	...	„ 65° 54′·3	... „ 7·6
„ (∵) „ „ 1647 „ 65°	...	„ 21h. 43m. 41s.	...	„ 65° 28′·4	... „ 8·1

The photograph was taken with the 20-inch reflector on September 25th, 1895, between sidereal time 20h. 48m. and 23h. 51m., with an exposure of the plate during three hours.

REFERENCES.

Neb. H IV. 75 N.G.C. 7129. G.C. 4702. *h* 2131.

Cl. H VII. 66 „ 7142. „ 4709. *h* 2134.

Sir J. Herschel, in the *Phil. Trans.* for 1833, No. 2131, p. 471, describes the object as a very coarse triple star involved in a nebulous atmosphere; a curious object. The nebula is exceedingly faint and graduates away—3 stars in the nebula.

Lord Rosse (*Obs. of Neb. and Cl.*, p. 162) describes the nebula as pretty bright; small; round; suddenly much brighter in the middle.

The photograph shows the cluster to be in agreement with Sir J. Herschel's descriptions, and of course accurately depicts each star down to about the 17th magnitude.

The nebula is shown to be elliptical, measuring 432″ in *n.f.* to *s.p.* direction, and 285″ in *s.f.* to *n.p.* The nebulosity is dense on the *n.f.* side, and involved in it, as a nucleus, are the three stars referred to by Sir J. Herschel; two of them are of about 12th mag., and the third about 16th mag. There are also eleven other stars, ranging between 12th and 17th mag., apparently involved in the nebulosity, the character of which is flocculent with extensive dark areas; and there is some structure visible in it near the *n.f.* margin.

There are three stars, each of about 13th mag., surrounded by faint nebulosity in the positions following:—measured from the centre of the tri-stellar nucleus of the nebula; (1) 358″ *north following*; (2) 326″ *north preceding*; (3) 277″ *north preceding.* The stars Nos. 2 and 3 are not referred to in the catalogues. The measurements given above are approximate.

PLATE IX.

Cluster H VI. 30 Cassiopeiæ.

R.A. 23h. 51m. 59s. Dec. N. 56° 9′·6.

The photograph covers the region between R.A. 23h. 49m. 42s. and R.A. 23h. 54m. 59s. Declination between 55° 37 ·6 and 56° 44′·2 North.

Scale—1 millimètre to 24 seconds of arc.

Co-ordinates of the Fiducial Stars marked with dots for the epoch A.D. 1900.

Star (.) D.M. No. 3045—Zone+55°	...	R.A. 23h. 50m. 23·6s.	...	Dec. N. 55° 52′·9	...	Mag. 8·9
,, (··) ,, ,, 3119 ,, 56°	...	,, 23h. 52m. 44·0s.	...	,, 56° 34′·4	...	,, 7·7

The photograph was taken with the 20-inch reflector on December 7th, 1898, between sidereal time 0h. 29m. and 1h. 59m., with an exposure of the plate during 90 minutes.

REFERENCES.

N.G.C. 7789. G.C. 5031. *h* 2284.

Sir J. Herschel (*Phil. Trans.*, 1833, p. 480) describes it as a most superb cluster, which fills the field and is full of stars; gradually brighter in the middle, but no condensation to a nucleus. Stars 11th to 18th magnitude.

Lord Rosse (*Obs of Neb. and Cl.*, p. 177) describes the cluster as very coarse, stars quite distinct and no nebulosity visible, dark holes and jagged branches, but no regular arrangement.

The photograph, together with the one taken six years ago and published in *I.R. Photos.*, Pl. 52, shows the cluster and the surrounding region to be densely studded with stars which, with few exceptions, are faint and arranged in very obvious curves, lines, lanes, and vacant spaces. These appear to me, after studying the many photographs of spiral nebulæ in this book and on the original negatives, to give evidence of their formation, or arrangement, by vortical movements similar in character to those shown in the spiral nebulæ. I have tried to detect, by superposition of the two negatives, if any indication of movements amongst the stars have taken place during the interval of six years, but such movements, being small in extent, will require careful measurements in order to determine them.

Plate 10.

NEBULA M. 31 ANDROMEDÆ.

NEBULA M. 33 TRIANGULI.

PLATE X.

Spiral Nebula M. 31 Andromedæ.

R.A. 0h. 37m. 17s. Dec. N. 40° 43′·4.

The photograph covers the region between R.A. 0h. 34m. 0s. and R.A. 0h. 40m. 41s. Declination between 39° 47′·2 and 41° 37′·2 North.

Scale—1 millimètre to 40 seconds of arc.

Co-ordinates of the Fiducial stars marked with dots for the epoch A.D. 1900.

Star (.)	D.M. No. 158—Zone+39°	... R.A. 0h. 36m. 35·5s.	... Dec. N. 40° 5′·5	... Mag. 7·0
„ (··)	„ „ 154 „ 40°	... „ 0h. 39m. 21·0s.	... „ 40° 45′·7	... „ 9·0
„ (·:·)	„ „ 158 „ 40°	... „ 0h. 40m. 33·6s.	... „ 40° 15′·6	... „ 7·5

The photograph was taken with the 20-inch reflector on October 17th, 1895, between sidereal time 23h. 19m. and 0h. 49m., with an exposure of the plate during ninety minutes.

REFERENCES.

N.G.C. 221, 224. G.C. 116, 117. *h* 50, 51. M 32, 31.

I.R. Photos., p. 31, where further references are given.

The photograph is one of a series taken between October 10th, 1887, and October 17th, 1895, and, although the plate was exposed during only ninety minutes, the nebulosity and the stars are depicted as densely and clearly as they are on other plates which were exposed during four hours; this may be due partly to the exceptional clearness of the atmosphere on the night when it was taken; similar unexpected results are of frequent occurrence in the practice of stellar photography.

That the nebula is a left-hand spiral, and not annular as I at first suspected, cannot now be questioned; for the convolutions can be traced up to the nucleus which resembles a small bright star at the centre of the dense surrounding nebulosity; but notwithstanding its density the divisions between the convolutions are plainly visible on negatives which have had a proper degree of exposure.

If we could view the nebula from a point perpendicular to its plane it would appear like some of the other spiral nebulæ which are depicted on plates 11 to 15 following, and its diameter would subtend an angle of about two and one-third degrees; but as we can only view it at an acute angle it has the appearance of an ellipse. This condition prevents us from seeing all the stellar condensations in the convolutions, but some of them are visible with sufficient distinctness to be available for the determination of relative movements between the nebula and stars within one degree distance. I have measured on two photographs, taken at an interval of eight years, the distances of two of the condensations from six stars around the nebula. The condensations are distant 81′·4 apart, and designated A and B. Of the six stars (a, b, c, d, e and f) four are on the *preceding* and two on the *following* side of the nebula, and they have been selected so as to form large angles with the condensations. The following are the results :—

	Measurements from A and B to stars on the *preceding* side of the nebula.				Measurements from A and B to stars on the *following* side of the nebula.	
Year 1888	c to B 1273″	d to B 2076″	e to A 1378″	f to A 2050″	a to A 1751″	b to B 2001″
,, 1896	,, 1268″	,, 2071″	,, 1377″	,, 2048″	,, 1748″	,, 2000″
	− ...5″	− ...5″	− ...1″	− ...2″	− ...3″	− ...1″

These measurements are within a probable error of one second of arc.

It would therefore appear that the nebula is very distant from the earth, for otherwise larger displacements than are here shown would have taken place within the interval of eight years ; and this shows that we must wait for the expiration of another similar or longer interval of time before these results can receive full confirmation ; but it should be noted that the differences have all the same (*minus*) sign, and this points to an approaching movement. There is also indication of rotation in the nebula, for the angle $c\ B\ d$, formed by the stars and condensation B, was twenty-one minutes of arc less in the year 1896 than in 1888 ; whilst the angle $e\ A\ f$, towards the other extremity of the nebula, had increased four minutes of arc during the same interval.

Since the question of variability in the nucleus of the nebula has been raised by several observers, the following evidence furnished by the photographs may be of interest :— A series of 42 photographs of the nebula were taken between January, 1891, and December, 1894, and on 26 of them the nucleus is not well defined ; but on 13 it is star-like. These facts do not absolutely prove that the nucleus is variable, for a little difference in the clearness of the atmosphere during the exposures of the plate might account for the apparent variability.

PLATE X.

Spiral Nebula M. 33 Trianguli.

R.A. 1h. 28m. 12s. Dec. N. 30° 8′·6.

The photograph covers the region between R.A. 1h. 26m. 36s. and R.A. 1h. 30m. 3s. Declination between 29° 37′·9 and 30° 43′·9 North.

Scale—1 millimètre to 24 seconds of arc.

Co-ordinates of the Fiducial Stars marked with dots for the epoch A.D. 1900.

Star (.) D.M. No. 260—Zone + 29°	...	R.A. 1h. 28m. 1·3s.	...	Dec. N. 29° 53′·6	... Mag. 8·0
„ (··) „ „ 265 „ 29°	...	„ 1h. 29m. 38·5s.	...	„ 30° 9′·3	. „ 9·2

The photograph was taken with the 20-inch reflector on November 14th, 1895, between sidereal time 1h. 18m. and 3h. 33m., with an exposure of the plate during two hours and fifteen minutes.

REFERENCES.

N.G.C. 598. G.C. 352. *h* 131. Rosse, *Obs. of Neb. and Cl. of Stars*, p. 20; *Phil. Trans.*, 1850, pl. 36, fig. 5, and 1861, pl. 36, fig. 10.

Sir J. Herschel, in the G.C., p. 352, describes the nebula as a remarkable object; extremely large; round; very rich; very gradually brighter in the middle with a nucleus; and resolvable.

Lord Rosse (references given above) has recorded 32 observations made between the years 1849 and 1862, and gives a marginal sketch.

The photograph shows the nebula to be a right hand spiral with remarkable features. It measures about 62 minutes of arc from *north following* to *south preceding*, and 35 minutes from *south following* to *north preceding*, and it is obvious that any attempt to delineate it by eye observations and drawing by hand could not possibly succeed owing to its vast area, faintness, and complicated structure. For similar reasons any attempt to describe it verbally would, with the photograph before us, be unnecessary. But there are some features clearly visible on the negatives that cannot be fully reproduced on the printed copies, and to those I will now refer.

There are twenty stars involved in the small patch of dense nebulosity which forms the central nuclear condensation, where the spirals meet, and each of the other patches of nebulosity seen scattered about has numerous faint nebulous stars involved, and they follow with much regularity the general trend of the convolutions. Where these are turned aside from the symmetrical development of their curvatures, the stars also are turned into loops, lines, and into various curves.

Involved in the general faint nebulosity, as well as in the various detached patches, are hundreds of stars having each a very small nucleus surrounded by nebulous matter; they are like nebulous stars in the midst of fainter nebulosity, and with few exceptions these stars would be classed as of the 16th to 17th magnitude.

It is by studying the original negatives that a full apprehension of the structure of this nebula can best be realized, and we should thereby infer that changes of a relatively rapid character must be taking place within it. But on comparing the negative of which the annexed photograph is a copy with another that was taken on the 11th of November, 1891, the interval being four years, I could not detect that any obvious change had taken place in any part of the nebula during that period; this therefore indicates that its distance from the earth is great.

The study of the photographs would justify us in suspecting that this nebula is the result of a celestial cataclysm such as the collision of two bodies like the sun, or of two dense swarms of meteorites, or of clouds of nebulous matter. Either of these suppositions might be tenable by the evidence, photographic and otherwise, which is now available to us, and photographs taken in future years will furnish the data necessary for demonstration.

Plate 11.

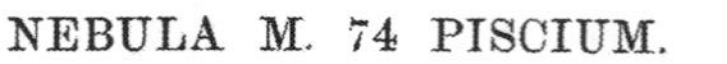

NEBULA M. 74 PISCIUM.

NEBULA H V. 44 CAMELOPARDI.

NEBULA H I. 285 URSÆ MAJORIS.

NEBULA H I. 199 URSÆ MAJORIS.

PLATE XI.

Spiral Nebula M. 74 Piscium.

R.A. 1h. 31m. 19s. Dec. N. 15° 16′·3.

The photograph covers 20′·4 in R.A. and 16′·3 in Declination, and was taken with the 20-inch reflector on December 9th, 1893, between sidereal time 1h. 7m. and 5h. 24m., with an exposure of the plate during three hours and forty minutes.

Scale—1 millimètre to 12 seconds of arc.

REFERENCES.

N.G.C. 628. G.C. 372. *h* 142.

Sir J. Herschel, in the G.C., states as the result of eleven observations that it is a globular cluster; faint; very large; round; pretty suddenly much brighter in the middle; partially resolved.

Lord Rosse, in the *Phil. Trans.* for 1861, p. 711, designates it as a spiral nebula with the centre formed of stars, and several stars visible through the nebula. A marginal sketch shows the spiral form, and in the *Obs. of Neb. and Cl.*, p. 21, another marginal sketch is given which shows a nucleus and five stars involved in the spirals.

The photograph shows the nebula to be a very perfect right-hand spiral, with a central stellar nucleus and a 15th mag. star close to it on the south side. The convolutions are studded with many stars and star-like condensations, and on the *north preceding* side there is a partial inversion of one of the convolutions, which conveys the idea of some disturbing cause having interfered with the regular formation of a part of that convolution.

It will be observed on examination of this, and the other photographs of spiral nebulæ, that the nebulous matter forming the convolutions is broken up into stars and star-like *loci*, and that they vary in brightness between the light of stars of about thirteenth and seventeenth magnitudes. The bright stars have as well-defined margins as any other stars in the sky, whilst the fainter ones are nebulous and not well defined round their margins; and those that differ but little in brightness from the nebulosity in which they are immersed have more or less undefined margins.

Every spiral nebula that I have photographed has a stellar nucleus surrounded by dense nebulosity in its centre of revolution, and around that centre the nebulous convolutions and the stars involved in them are more or less symmetrically arranged. The records of these features are now so numerous and accordant that they cannot be attributed to accidental or to fortuitous circumstances.

PLATE XI.

Spiral Nebula H V. 44 Camelopardi.

R.A. 7h. 27m. 9s. Dec. N. 65° 48′·9.

The photograph covers 20′·4 in R.A. and 16′·3 in Declination, and was taken with the 20-inch reflector on March 21st, 1898, between sidereal time 7h. 37m. and 9h. 27m., with an exposure of the plate during ninety minutes.

Scale—1 millimètre to 12 seconds of arc.

REFERENCES.

N.G.C. 2403. G.C. 1541.

Sir J. Herschel, in the G.C., describes it as a very remarkable object; considerably bright; extremely large; very much extended; gradually much brighter in the middle and with a nucleus.

The photograph shows the nebula to be a right hand spiral, with several bright stars, besides numerous stellar condensations apparently involved in the convolutions. The nucleus is very dense, and seems to consist of more than one star, and the nebula generally resembles N. 33 *Trianguli*. The major diameter extends in *n.p.* to *s.f.* direction, and can, on the negative, be traced to about 18′ of arc in extent, but much of the external nebulosity is too faint for reproduction in print. The nebula N.G.C. 2404 (*Bigourdan*) is one of the numerous star-like nebulous condensations which are shown by the photograph surrounding this nebula; there are many such within a distance of one degree from the nucleus.

PLATE XI.

Spiral Nebula H I. 285 Ursæ Majoris.

R.A. 9h. 38m. 58s. Dec. N. 68° 22′·7.

The photograph covers 20′·4 in R.A. and 16′·3 in Declination, and was taken with the 20-inch reflector on March 28th, 1895, between sidereal time 9h. 9m. and 10h. 39m., with an exposure of the plate during ninety minutes.

Scale—1 millimètre to 12 seconds of arc.

REFERENCES.

N.G.C. 2976. G.C. 1905. *h* 625.

Sir J. Herschel, in the G.C., describes the nebula as bright; very large; much extended, 152°; stars involved.

The photograph shows the nebula to be probably a left-hand spiral, with the greatest elongation in *s.f.* to *n.p.* direction. There are six faint star-like condensations involved besides other irregular condensations distributed over the nebula.

PLATE XI.

Spiral Nebula H I. 199 Ursæ Majoris.

R.A. 10h. 13m. 42s. Dec. N. 46° 3′·7.

The photograph covers 20′·4 in R.A. and 16′·3 in Declination, and was taken with the 20-inch reflector on April 17th, 1898, between sidereal time 10h. 47m. and 13h. 7m., with an exposure of the plate during two hours and twenty minutes.

Scale—1 millimètre to 12 seconds of arc.

REFERENCES.

N.G.C. 3198. G.C. 2066. *h* 695.

Sir J. Herschel, in the G.C., describes the nebula as pretty bright; very large; much extended 45° ±; very gradually brighter in the middle.

Lord Rosse (*Obs of Neb. and Cl.*, p. 83) describes it as probably a faint spiral; star in *f.* end; light very unequal; much extended; suddenly brighter in the middle; with a round nucleus; dark spaces throughout its length.

The photograph shows the nebula to be a left-hand spiral with elongated stellar nucleus and a star of about 17th magnitude involved on the *n.f.* side, and there are several stellar condensations visible in the convolutions. The major axis extends in *n.f.* to *s.p.* direction.

Plate 12.

NEBULA M. 65 LEONIS.

NEBULA H IV. 6 SEXTANTIS.

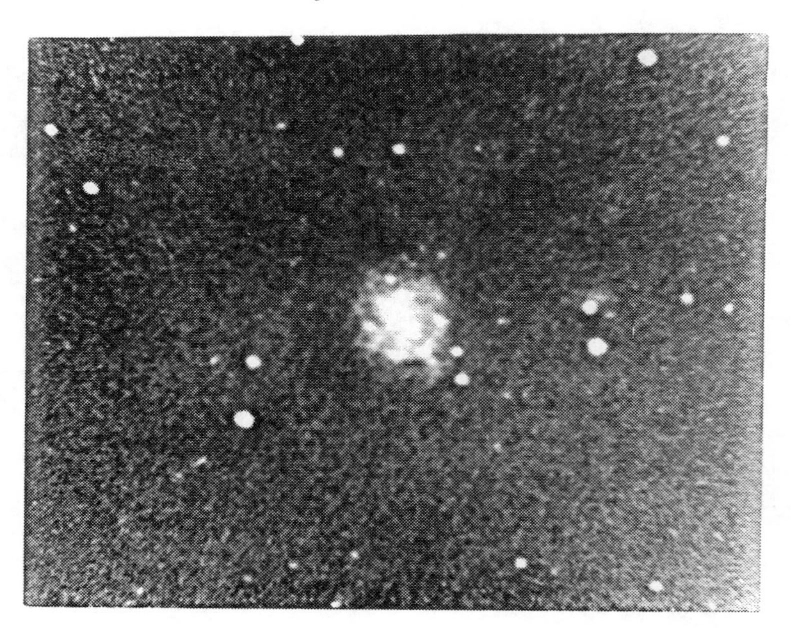

NEBULA M. 66 LEONIS.

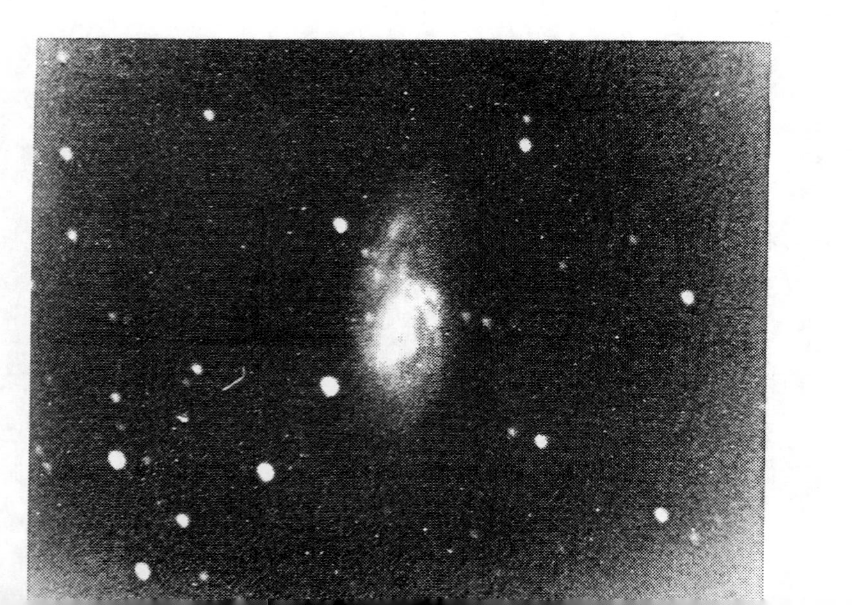

NEBULA H I. 226 URSÆ MAJORIS.

PLATE XII.

Spiral Nebula H IV. 6 Sextantis.

R.A. 10h. 46m. 3s. Dec. N. 6° 22′·3.

The photograph covers 20′·4 in R.A. and 16′·3 in Declination, and was taken with the 20-inch reflector on April 22nd, 1898, between sidereal time 10h. 52m. and 13h. 43m., with an exposure of the plate during two hours and fifty-one minutes.

Scale—1 millimètre to 12 seconds of arc.

REFERENCES.

N.G.C. 3423. G.C. 2234. *h* 777.

Sir J. Herschel, in the G.C., describes the nebula as faint; very large; round; very gradually brighter in the middle; partly resolvable.

Lord Rosse (*Obs. of Neb. and Cl.*, p. 89) describes it as a pretty bright central disc in a ring of nebulosity, and a drawing of it is given on Pl. III., No. 3.

The photograph shows the nebula to be a left-hand spiral with a faint stellar nucleus surrounded by nebulosity. There is a 17th magnitude star *s.p.* the nucleus and 11 stellar condensations in the convolutions.

The granulation of the film of the negative, which is prominently shown on this as well as on other photographs, must not be mistaken for nebulosity, or stars; this appearance is unavoidable on enlargements above five times the original, but real star-images and nebulosity are easily distinguishable.

PLATE XII.

Spiral Nebula M. 65 Leonis.

R.A. 11h. 13m. 43s. Dec. N. 13° 38′·5.

The photograph covers 20′·4 in R.A. and 16′·3 in Declination, and was taken with the 20-inch reflector on February 28th, 1894, between sidereal time 8h. 19m. and 12h. 23m., with an exposure of the plate during three hours and forty minutes.

Scale—1 millimètre to 12 seconds of arc.

REFERENCES.

N.G.C. 3623. G.C. 2373. *h* 854.

Sir J. Herschel describes the nebula in the G.C., and in the *Phil. Trans.* for 1833, pl. XIV., fig. 53, is a drawing of it.

Lord Rosse, in the *Phil. Trans.*, 1850, pl. XXXVII., fig. 7, gives a drawing of the nebula, and in the *Obs. of Neb. and Cl.*, p. 95, gives the results of eight observations upon it.

The photograph shows the nebula to be a left-hand spiral, with the external outline so regularly formed that it resembles an annular nebula with rings encircling it ; but the spiral form must be the true interpretation, and the rings of nebulosity with the dark spaces between them and the nebulous star-like condensations together form parts of the convolutions ; the dark spaces being the intervals between them.

PLATE XII.

Spiral Nebula M. 66 Leonis.

R.A. 11h. 15m. 2s. Dec. N. 13° 32′·4.

The photograph covers 20′·4 in R.A. and 16′·3 in Declination, and was taken with the 20-inch reflector on February 28th, 1894, between sidereal time 8h. 19m. and 12h. 23m., with an exposure of the plate during three hours and forty minutes.

Scale—1 millimètre to 12 seconds of arc.

REFERENCES.

N.G.C. 3627. G.C. 2377. *h* 857.

Sir J. Herschel, in the G.C., gives a description of the nebula and a drawing in the *Phil. Trans.*, 1833, pl. XIV., fig. 54.

Lord Rosse, in the *Obs. of Neb. and Cl.*, p. 95, describes the nebula as a spiral, and a marginal sketch is given. In the *Phil. Trans.*, 1861, pl. XXXVII., fig. 16, is a drawing of it.

The photograph shows the nebula to be a spiral, with a well defined stellar nucleus which forms the pole of the convolutions in which I have counted fourteen nebulous star-like condensations.

It must be understood in every case where these small faint nebulæ are described in this book that the particulars given are based on the *original negatives*, for the details cannot all be copied on the paper enlargements. In case of any doubt the negatives can be referred to.

PLATE XII.

Spiral Nebula H I. 226 Ursæ Majoris.

R.A. 11h. 15m. 24s. Dec. N. 53° 43′·8.

The photograph covers 20′·4 in R.A. and 16′·3 in Declination, and was taken with the 20-inch reflector on April 29th, 1896, between sidereal time 11h. 20m. and 14h. 9m., with an exposure of the plate during two hours and forty-nine minutes.

Scale—1 millimètre to 12 seconds of arc.

REFERENCES.

N.G.C. 3631. G.C. 2379. *h* 858.

Sir J. Herschel, in the G.C., describes the nebula as pretty bright; large; round; suddenly very much brighter in the middle; mottled and with a nucleus.

Lord Rosse (*Obs. of Neb. and Cl.*, p. 96) describes it as a spiral, with two arms which are broken and of unequal light; there are bright patches in it.

The photograph shows the nebula to be a right-hand spiral, with a bright stellar nucleus; the convolutions are nearly symmetrical, and immersed in them are numerous star-like condensations. The stars in the surrounding region are few in number, and those shown on the accompanying plate, though they are slightly elongated, will serve as points of reference with regard to possible changes that in future may take place in the nebula.

Plate 13.

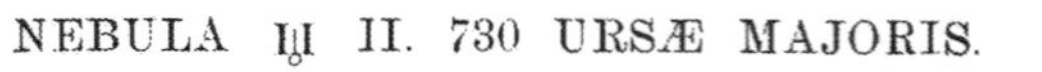

NEBULA H II. 730 URSÆ MAJORIS.

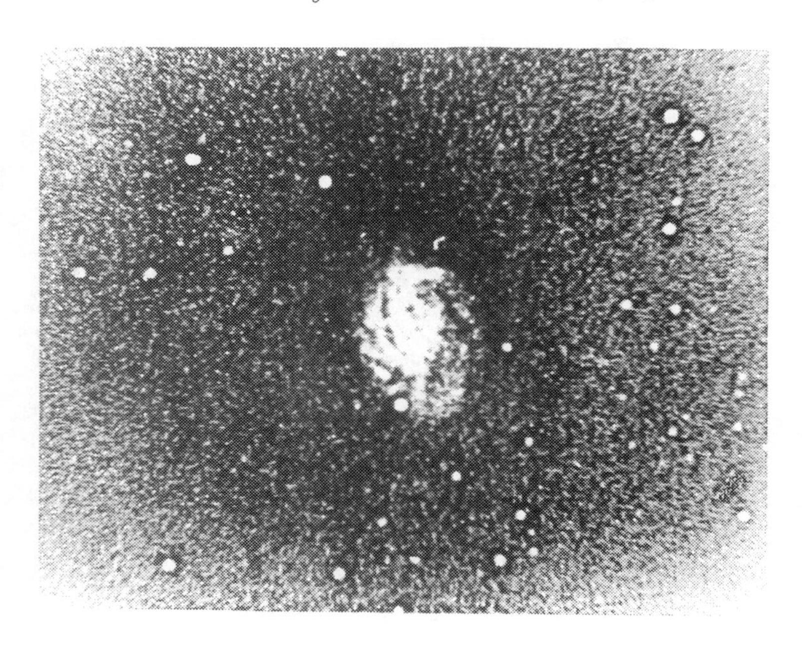

NEBULA H I. 206 URSÆ MAJORIS.

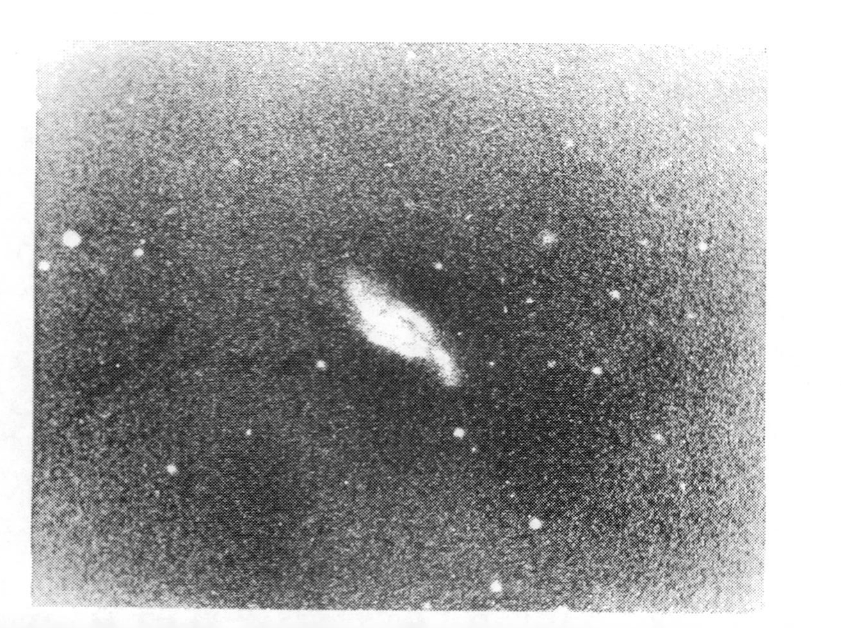

NEBULA H IV. 56 URSÆ MAJORIS.

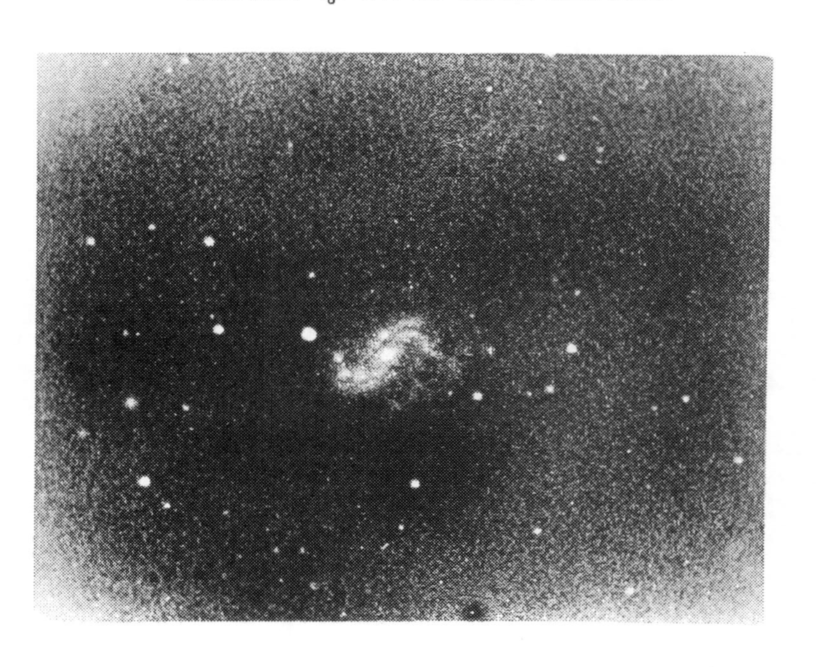

NEBULA M. 99 VIRGINIS.

PLATE XIII.

Spiral Nebula H̥ II. 730 Ursæ Majoris.

R.A. 11h. 27m. 56s. Dec. N. 47° 35′·8.

The photograph covers 20′·4 in R.A. and 16′·3 in Declination, and was taken with the 20-inch reflector on April 12th, 1898, between sidereal time 10h. 26m. and 11h. 56m., with an exposure of the plate during ninety minutes.

Scale—1 millimètre to 12 seconds of arc.

REFERENCES.

N.G.C. 3726. G.C. 2445. *h* 910.

Sir J. Herschel, in the G.C., describes the nebula as pretty bright; very large; little extended in direction 0°; very gradually much brighter in the middle; with star 15 mag.

Lord Rosse (*Obs. of Neb. and Cl.*, p. 98) observed the nebula between 1848 and 1861, and glimpsed traces of spirality in it, and also saw some of the brighter parts of the nebulous arms.

The photograph shows the nebula to be a left-hand spiral, with a bright stellar nucleus and many star-like condensations in the convolutions.

The coarseness of the film shown on some of the photographs, as well as the lighter parts about the extreme corners of the plates, are due to instrumental causes in the process of enlargement and not to stars or nebulosity; the stars are readily distinguishable from film defects, on all the photographs, by the character of their discs.

PLATE XIII.

Spiral Nebula H IV. 56 Ursæ Majoris.

R.A. 11h. 58m. 3s. Dec. N. 45° 5′·0.

The photograph covers 20′·4 in R.A. and 16′·3 in Declination, and was taken with the 20-inch reflector on May 3rd, 1897, between sidereal time 11h. 57m. and 13h. 27m., with an exposure of the plate during ninety minutes.

Scale—1 millimètre to 12 seconds of arc.

REFERENCES.

N.G.C. 4051. G.C. 2680. *h* 1061.

Sir J. Herschel, in the G.C., describes the nebula as bright; very large; extended; very gradually, very suddenly much brighter in the middle; star 11 mag.

Lord Rosse (*Obs. of Neb. and Cl.*, p. 107) made nine observations between 1851 and 1862, and found it to be a spiral; light mottled faint nebulosity fills up the spaces between the arms. Figured in the *Phil. Trans.*, 1861, pl. XXVII., 19.

The photograph shows the nebula to be a left-hand spiral, with bright stellar nucleus, and with both stellar and elongated condensations in the convolutions.

PLATE XIII.

Spiral Nebula H I. 206 Ursæ Majoris.

R.A. 12h. 0m. 28s. Dec. N. 51° 5 ·9.

The photograph covers 20′·4 in R.A. and 16′·3 in Declination, and was taken with the 20-inch reflector on May 3rd, 1896, between sidereal time 11h. 59m. and 14h. 49m., with an exposure of the plate during two hours and fifty minutes.

Scale—1 millimètre to 12 seconds of arc.

REFERENCES.

N.G.C. 4088. G.C. 2708.

Sir J. Herschel, in the G.C., describes the nebula as bright; considerably large; pretty much extended 135° ± ; little brighter in the middle.

Lord Rosse (*Obs. of Neb. and Cl.*, p. 108) describes it as very large, extended 53°·5, a new spiral, with probably many details of interest; and that it is of an S shape, with dark lanes in it. Four observations were made between 1867 and 1873.

The photograph shows the nebula to be a left-hand spiral with three prominent convolutions and several stellar condensations in them; the nucleus also is stellar.

PLATE XIII.

Spiral Nebula M. 99 Virginis.

R.A. 12h. 13m. 45s. Dec. N. 14° 58′·5.

The photograph covers 20′·4 in R.A. and 16′·3 in Declination, and was taken with the 20-inch reflector on May 4th, 1896, between sidereal time 12h. 19m. and 15h. 16m., with an exposure of the plate during two hours and fifty-seven minutes.

Scale—1 millimètre to 12 seconds of arc.

REFERENCES.

N.G.C. 4254. G.C. 2838. *h* 1173.

Sir J. Herschel, in the G.C., describes it as a very remarkable object; bright; large; round; gradually brighter in the middle; mottled.

Lord Rosse (*Obs. of Neb. and Cl.*, p. 112) made five observations of it between 1848 and 1861 and recognised its spiral character. A drawing of it is given in the *Phil. Trans.*, 1850, pl. XXXV., fig. 2.

Lassell, in the *Mem. R.A.S.*, Vol. 36, pl. IV., fig. 16, gives a drawing of it.

The photograph shows the nebula to be a right-hand spiral, with a composite stellar nucleus and many star-like condensations in the convolutions. The following are the measurements of the Position Angles and distances, from the centre of the composite nucleus, of the condensations, and of the stars which are within a distance of 470 seconds of arc from it; the diagram (plate XVI.) will also be of assistance in the recognition of the objects measured. The stars are designated with numbers, and the condensations with letters of the alphabet.

TABLE OF MEASUREMENTS.—Epoch 1896.

(*See* diagram, Plate No. XVI.)

Designation of Stars and Condensations.	Position Angles.			Distance from Nucleus.	Designation of Stars and Condensations.	Position Angles.			Distance from Nucleus.
1.	6°	26′	7″	320″	7.	127°	13′	5″	353″
a.	16	8	8	112	m.	166	10	36	45
b.	19	50	0	124	n.	196	18	8	48
2.	31	28	18	291	8.	207	17	35	291
A. 3.	34	30	11	186	o.	208	7	45	89
c.	35	56	35	128	p.	209	8	5	75
4.	37	2	55	273	q.	230	22	58	63
d.	42	25	58	71	9.	264	16	0	371
e.	54	2	2	22	10.	266	25	23	468
f.	61	43	22	121	r.	278	30	0	113
g.	78	36	43	42	s.	294	17	3	136
B. h.	78	36	43	113	t.	312	55	23	173
5.	82	56	16	300	11.	321	20	0	361
i.	99	11	27	83	u.	334	41	57	52
k.	99	11	27	136	v.	345	55	0	94
C. 6.	114	45	0	115	w.	352	10	38	70
l.	117	17	6	71					

Measurements of the Position Angles and distances from the nucleus (N) of the nebula, of the three stars A, B and C, by Lord Rosse in the year 1849 (*Obs. of Neb. and Cl. of St.*, p. 112), and also of the corresponding photographic distances.

	Rosse.		Photograph.	
N A	Pos. 33°	Dist. 175″	Pos. 34° 30′ 11″	Dist. 186″
N B	„ 80	„ 106	„ 78 36 43	„ 113
N C	„ 117	„ 108	„ 114 45 0	„ 115
N D	„ 177	„ 168	Not on the photograph.	

The star D cannot be recognised on the photograph, and it will be observed that the Position Angles and distances from the nucleus of the stars A, B and C indicate that changes have taken place amongst them during the interval between the years 1849 and 1896 ; but astronomers must wait some years longer for other photographs to be taken and measured before definitely pronouncing judgment upon these apparent changes. I may, however, remark that the increase in the distances from the nucleus of the stars A, B and C are consistent with a movement of recession of the nebula along the line of sight, to the equivalent distance of about nine seconds of arc measured from these stars. There are also differences in the Position Angles which are consistent with a movement of rotation of the nebula upon the nucleus as axis.

The measurements of the stars and star-like condensations which are given in the Table have been made upon the negative taken in 1896, and the distances of two of the condensations, designated h and q, have been measured from stars so distant from the nebula that they are probably not physically connected with it. They are Fiducial stars available to detect changes in the nebula either of rotation or translation, and are respectively numbered in the Table and on the diagram 1, 7, 8, 9 and 11.

A third photograph of the nebula was taken in 1899, and in addition to the condensations given in the Table there are ten others, of which the Position Angles and distances from the nucleus, epoch 1899, are as follows :—

Pos. Angles.		Distance.		Pos. Angles.		Distance.
13° 56′	...	38″	......	110° 30′	...	49″
29 0	...	134	......	119 45	...	19
31 30	...	63	......	248 45	...	104
88 51	...	73	......	266 38	...	37
104 6	...	60	......	298 10	...	48

The condensation marked b in the Table, Pos. Angle 19° 50′ and distance 124″, is missing on the 1899 photograph ; whereas all the others are shown upon it and their measurements given.

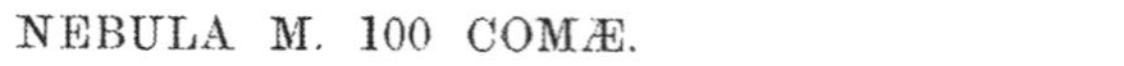
NEBULA M. 100 COMÆ.

NEBULA ♅ V. 2 VIRGINIS.

NEBULA ♅ I. 92 COMÆ.

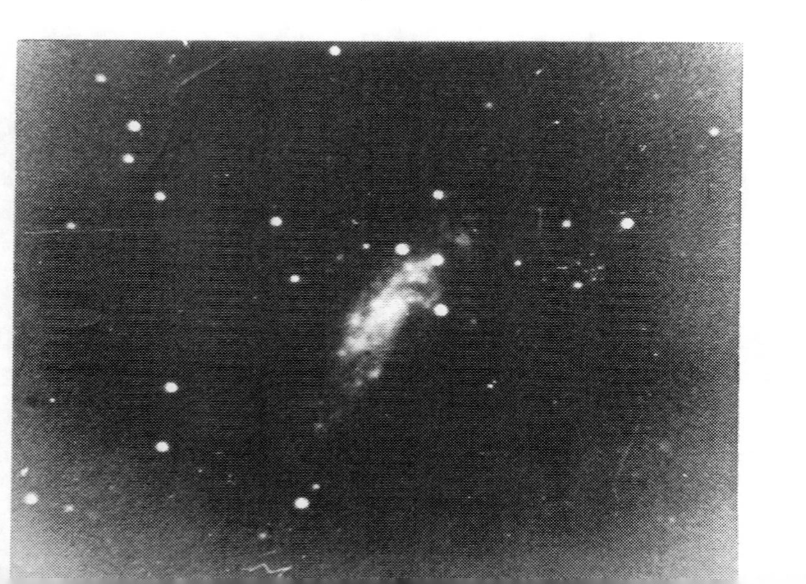

NEBULA ♅ I. 84 COMÆ.

PLATE XIV.

Spiral Nebula M. 100 Comæ Berenicis.

R.A. 12h. 17m. 52s. Dec. N. 16° 22′·7.

The photograph covers 20′·4 in R.A. and 16′·3 in Declination, and was taken with the 20-inch reflector on May 9th, 1896, between sidereal time 12h. 38m. and 15h. 33m., with an exposure of the plate during two hours and fifty-five minutes.

Scale—1 millimètre to 12 seconds of arc.

REFERENCES.

N.G.C. 4321. G.C. 2890. *h* 1211.

Sir J. Herschel, in the G.C., describes it as a remarkable object; pretty faint; very large; round; very gradually, then very suddenly brighter in the middle; mottled nucleus.

Lord Rosse (*Obs. of Neb. and Cl.*, p. 113) describes it as a right-handed spiral, very faint, very large, the central part is a beautiful planetary nebula and seems almost detached from the surrounding faint nebulosity.

Lassell, in the *Mem. of the R.A.S.*, Vol. XXXVI., p. 44, gives a drawing (pl. III., fig. 7), and states that the spiral form is forcibly suggested.

The photograph shows the nebula to be a left-hand spiral with the nucleus very sharply stellar in the midst of faint nebulosity. The convolutions are strikingly perfect, and have several aggregations of nebulosity in them; three or four faint stars are also involved. Three faint small nebulæ are shown on the plate; one of them is ℍ II. 628, on the *s.p.* side; another is N.G.C. 4322, on the *n.f.* side; the third is ℍ II. 84, on the *f.* side. There are also two other nebulæ between the two last named, and fainter than them, that are not referred to in the catalogues.

PLATE XIV.

Spiral Nebula H V. 2 Virginis.

R.A. 12h. 29m. 20s. Dec. N. 2° 44′·2.

The photograph covers 20′·4 in R.A. and 16′·3 in Declination, and was taken with the 20-inch reflector on March 25th, 1894, between sidereal time 8h. 53m. and 11h. 56m., with an exposure of the plate during three hours.

Scale—1 millimètre to 12 seconds of arc.

REFERENCES.

N.G.C. 4536. G.C. 3085. *h* 1337.

Sir J. Herschel, in the G.C., describes the nebula as bright; very large; much extended, 110°; suddenly brighter in the middle; extremely mottled.

Lord Rosse (*Obs. of Neb. and Cl.*, p. 117) saw it as represented in the drawing in the *Phil. Trans.*, 1861, pl. XXVIII., fig. 24.

The photograph shows the nebula to be a right-hand spiral, viewed obliquely; it has a bright stellar nucleus, in the midst of dense nebulosity and several star-like condensations, and irregularly formed aggregations of nebulosity in the convolutions. On the north edge of the plate is shown a part of the nebula N.G.C. 4533.

The black spots and the granulations seen on the plate are due to defects in the film of the negative.

PLATE XIV.

Spiral Nebula H I. 92 Comæ Berenicis.

R.A. 12h. 31m. 0s. Dec. N. 28° 30′·6.

The photograph covers 20′·4 in R.A. and 16′·3 in Declination, and was taken with the 20-inch reflector on May 1st, 1894, between sidereal time 11h. 46m. and 13h. 46m., with an exposure of the plate during two hours.

Scale—1 millimètre to 12 seconds of arc.

REFERENCES.

N.G.C. 4559. G.C. 3101. *h* 1352.

Sir J. Herschel, in the G.C., describes the nebula as very bright; very large; much extended, 150°; gradually brighter in the middle; three stars following. A drawing is given in the *Phil. Trans.*, 1833, pl. XVI., fig. 83.

Lord Rosse (*Obs. of Neb. and Cl.*, p, 118) describes it as a very large ray, gradually brighter in the middle; some stars involved; edges irregularly defined; three bright stars at the *f.* end.

The photograph shows the nebula to be a left hand spiral, viewed obliquely. It has a small bright stellar nucleus, surrounded by dense nebulosity; several stars and star-like condensations are involved in it. The three bright stars on the *s.f.* end are just clear of the nebulosity, and the faint detached wisps of nebulous matter on the *s.p.* and *n.p.* and *s.f.* sides probably form parts of the nebula. There is a peculiar ring of nebulous matter shown close to the *s.f.* side of the nucleus that lies across one of the convolutions, which it seems to have deformed; and they are rather irregular in outline.

PLATE XIV.

Spiral Nebula H I. 84 Comæ Berenicis.

R.A. 12h. 45m. 33s. Dec. N. 26° 2′·7.

The photograph covers 20′·4 in R.A. and 16′·3 in Declination, and was taken with the 20-inch reflector on May 7th, 1894, between sidereal time 12h. 31m. and 14h. 1m., with an exposure of the plate during ninety minutes.

Scale—1 millimètre to 12 seconds of arc.

REFERENCES.

N.G.C. 4725. G.C. 3249. *h* 1451.

Sir J. Herschel, in the G.C., describes the nebula as very bright; very large; very gradually, then very suddenly, very much brighter in the middle; extremely bright nucleus.

Lord Rosse (*Obs. of Neb. and Cl., p.* 121) describes it as another spiral with two arms and some stars in the *f.* arm; the centre is bright. The centre itself is like an extended nebula with a nucleus.

The photograph shows the nebula with a large bright stellar nucleus surrounded by convolutions of nebulous matter with several stars and star-like condensations in them; there are three stars of about 14th mag. and three of about 16th mag. between one of the convolutions and the nucleus.

Plate 15.

NEBULA M. 64 COMÆ.

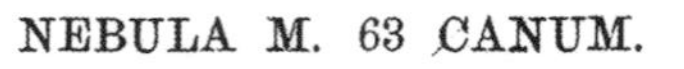

NEBULA M. 63 CANUM.

NEBULA M. 51 CANUM.

NEBULA H IV. 76 CEPHEI.

PLATE XV.

Spiral Nebula M. 64 Comæ Berenicis.

R.A. 12h. 51m. 49s. Dec. N. 22° 13′·9.

The photograph covers 20′·4 in R.A. and 16′·3 in Declination, and was taken with the 20-inch reflector on May 10th, 1896, between sidereal time 12h. 48m. and 15h. 44m., with an exposure of the plate during two hours and fifty-six minutes.

Scale—1 millimètre to 12 seconds of arc.

REFERENCES.

N.G.C. 4826. G.C. 3321. *h* 1486.

Sir J. Herschel, in the *Phil. Trans.*, 1833, p. 436, describes the nebula as bright; very large; very much extended; very suddenly much brighter in the middle; nucleus supposed to be a double star; nebula is 8′ or 9′ long, and 2′ broad; oval; with a vacuity between the nucleus. A drawing is given on pl. X., fig. 27.

Lord Rosse (*Obs. of Neb. and Cl.*, p. 123) describes it as curious, circular shape, with a dark large spot on one side, around which is a close cluster of well-defined little stars. Symptoms of resolvability strongly marked, particularly about the black "eye."

Lassell, in the *Mem. R.A.S.*, Vol. XXXVI., p. 46, describes it, and gives a drawing, pl. VI., fig. 26.

The photograph shows the nebula to be a right-hand spiral with a large bright stellar nucleus; one of the convolutions on the *n.f.* side of the nucleus is very bright, with a dark space, between it and the nucleus, which is free from nebulosity—thus producing the effect of contrast between light and dark areas. There are at least six very faint stars involved in the nebulosity besides some star-like condensations.

PLATE XV.

Spiral Nebula M. 63 Canum Venaticorum.

R.A. 13h. 11m. 20s. Dec. N. 42° 33′·6.

The photograph covers 20′·4 in R.A. and 16′·3 in Declination, and was taken with the 20-inch reflector on May 14th, 1896, between sidereal time 13h. 13m. and 16h. 8m., with an exposure of the plate during two hours and fifty-five minutes.

Scale—1 millimètre to 12 seconds of arc.

REFERENCES.

N.G.C. 5055. G.C. 3474. *h* 1570.

Sir J. Herschel, in the G.C., describes the nebula as very bright; large; pretty much extended, 120° ±; very suddenly much brighter in the middle; bright nucleus.

Lord Rosse (*Obs. of Neb. and Cl.*, p. 126) seemed uncertain about its spiral character.

The photograph shows the nebula to be a right hand spiral, with a bright stellar nucleus in the centre of dense nebulosity. The nebula is viewed rather obliquely, but the convolutions appear to be very regular in form, and studded with a large number of faint stars and stellar condensations.

PLATE XV.

Spiral Nebula H IV. 76 Cephei.

R.A. 20h. 32m. 48s. Dec. N. 59° 48′·0.

The photograph covers 20′·4 in R.A. and 16′·3 in Declination, and was taken with the 20-inch reflector on October 9th, 1896, between sidereal time 20h. 36m. and 23h. 31m., with an exposure of the plate during two hours and fifty-five minutes.

Scale—1 millimètre to 12 seconds of arc.

REFERENCES.

N.G.C. 6946. G.C. 4594. *h* 2084.

Sir J. Herschel, in the G.C., describes the nebula as very faint ; very large ; very gradually, then very suddenly brighter in the middle ; partly resolvable.

Lord Rosse (*Obs. of Neb. and Cl.*, p. 157) describes it as a new spiral, very fine but faint, three branches, of which two terminate in knots, a fourth branch *n.p.* is very doubtful. A drawing of it is given in the *Phil. Trans.*, 1861, pl. XXX., fig. 36.

The photograph shows the nebula to be a right-hand spiral, with a stellar nucleus ; the convolutions are studded with stars, and star-like condensations involved in nebulosity. There is also faint nebulosity between the convolutions as well as bright stars apparently involved in them. This nebula is in a part of the sky where the stars are faint but very numerous.

PLATE XV.

Spiral Nebula M. 51 Canum Venaticorum.

R.A. 13h. 25m. 40s. Dec. N. 47° 42′·7.

The photograph covers 20′·4 in R.A. and 16′·3 in Declination, and was taken with the 20-inch reflector on April 15th, 1898, between sidereal time 12h. 15m. and 13h. 45m., with an exposure of the plate during ninety minutes.

Scale—1 millimètre to 12 seconds of arc.

REFERENCES.

N.G.C. 5194, 5195. G.C. 3572, 3574. *h* 1622, 1623. H I. 196.

I.R. Photos., p. 85, pl. XXX.

Sir J. Herschel, in the G.C., describes this as two nebulæ, the first, a magnificent object, with a nucleus and a ring. The other bright; pretty small; round; very gradually brighter in the middle. In the *Phil. Trans.*, 1833, pp. 496, 497, a detailed account is given, together with a drawing, pl. X., fig. 25.

Lord Rosse (*Phil. Trans.*, 1850, pp. 504, 505, pl. XXXV., fig. 1) describes the nebula, and shows in it a strong spiral structure. In his *Obs. of Neb. and Cl.*, pp. 127 to 131, and pl. IV., the nebula is very fully described and illustrated.

Lassell (*Mem. R.A.S.*, Vol. XXXVI., pp. 46, 47, pl. VI., figs. 27, 27a) describes and delineates the nebula.

The photograph shows the nebula to be a left-hand spiral as described by Lord Rosse; and it has been observed, by him and by others, more frequently and closely than any other of the objects of its class.

The nucleus of the large nebula consists of a small bright star in the midst of a patch of very dense nebulosity, from which the convolutions radiate in approximately symmetrical forms. The convolutions are broken up into numerous stars and star-like condensations, and there are wisps of nebulosity, with a star involved in each of them, projecting from some of the convolutions; one of these appears to have been deformed, for we see disarrangement of symmetry; there are also indications that this may have been caused by the action of the second nucleus, for it is apparently connected with the distorted part. There are some well-formed bright stars apparently placed in the spaces between the convolutions, which spaces are also filled with faint nebulosity.

The nucleus of the small nebula on the north side of the large one consists of a bright stellar centre, slightly elongated, in the midst of dense nebulosity; and there is also a faint nebulosity extending around it to a considerable distance on the north, south, and *preceding* sides.

Reference should be made to pages 24 to 26 *ante*, where the measurements of the position angles, and distances, from the nuclei, of the stars and condensations are given from the photograph, in juxtaposition with those recorded by Lord Rosse, from which inferences with regard to movements in the stars and condensations since the year 1851 are discussed and commented upon.

DIAGRAM OF M. 99 VIRGINIS.

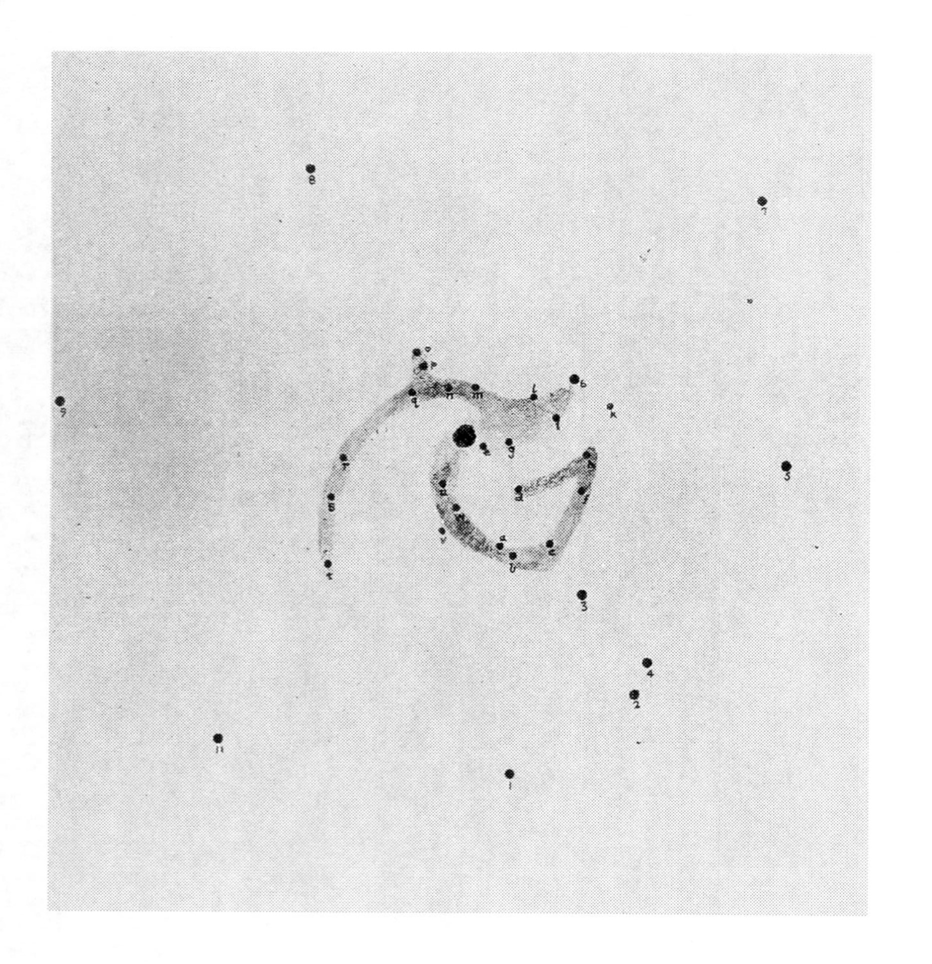

DIAGRAM OF M. 51 CANUM.

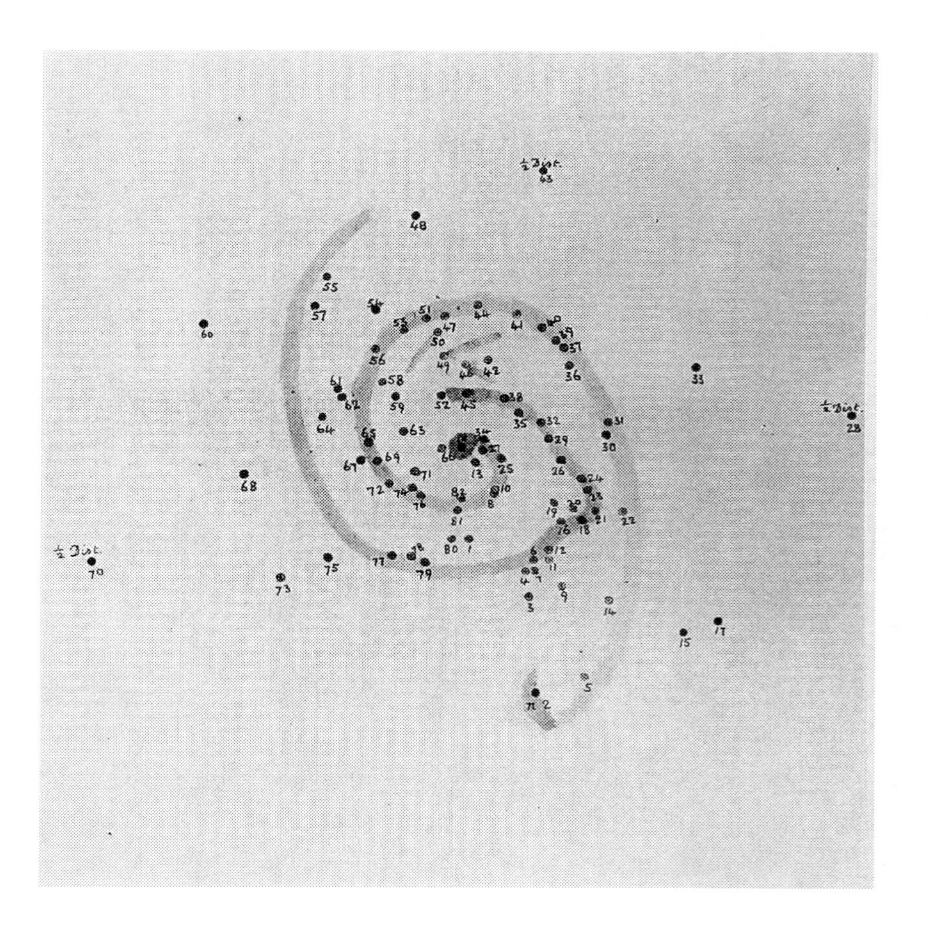

TABLE

Of the measurements of the Position Angles and distances from the nucleus (N) of the stars and condensations in and around the nebula. See Photograph (Plate XV.) and Diagram (Plate XVI.).

Number.	Position Angle.			Distance from the Nucleus.	Number.	Position Angle.			Distance from the Nucleus.
1	3°	49′	41″	99″	42	163°	58′	46″	100″
2	16	52	10	261	43	164	56	48	591
3	23	22	53	158	44	174	44	1	156
4	27	8	32	142	45	175	28	10	64
5	28	7	44	257	46	178	50	46	85
6	29	57	44	153	47	188	8	28	145
7	33	21	35	136	48	190	51	45	242
8	35	25	38	67	49	191	46	9	95
9	36	0	0	174	50	191	46	9	121
10	37	47	16	55	51	194	25	47	141
11	37	47	16	154	52	200	47	40	61
12	40	6	22	138	53	205	6	2	136
13	42	7	8	30	54	211	38	19	170
14	43	45	0	212	55	217	45	0	225
15	50	4	1	290	56	220	20	0	132
16	53	2	47	120	57	224	39	29	204
17	55	49	34	314	58	228	50	0	116
18	58	31	10	144	59	231	12	5	85
19	59	2	51	115	60	242	42	41	284
20	61	11	5	124	61	243	18	22	146
21	64	12	18	155	62	246	29	31	134
22	68	30	24	172	63	253	40	0	69
23	70	54	22	133	64	257	8	9	147
24	75	13	40	121	65	267	22	40	95
25	75	26	31	42	66	273	12	48	29
26	83	16	25	102	67	277	29	27	226
27	86	18	8	27	68	277	29	27	111
28	94	37	44	760	69	281	17	4	85
29	94	56	48	86	70	288	14	21	770
30	94	56	48	147	71	297	8	19	59
31	99	47	32	155	72	297	8	19	81
32	108	25	48	79	73	307	5	0	233
33	109	32	3	241	74	308	50	51	78
34	113	30	0	31	75	310	6	43	181
35	122	20	0	76	76	318	18	59	76
36	127	51	50	140	77	328	16	28	131
37	136	8	30	150	78	336	0	0	123
38	139	44	17	75	79	342	12	52	120
39	139	44	17	151	80	353	37	17	109
40	148	1	46	156	81	357	18	16	78
41	158	56	15	159	82	358	4	22	64

The stars numbered 28, 43, and 70 are marked on the Diagram Plate with ½ *Dist.*, which means that they are placed, for convenience, at half their measured distances by scale from the nucleus.

It has already been stated in the introduction, under the heading, "THE ARRANGEMENT OF THE PHOTOGRAPHS," that the edge of the plate next to the printed heading is the *south*, and the opposite edge the *north;* the meridian for the measurement of position angles is therefore drawn from north to south through the nucleus "N," and the position angles measured in the ordinary way from this meridian, as shown by the sequence of the figures.

Plate 17.

NEBULA H V. 46 URSÆ MAJORIS.

NEBULÆ H I. 197-8 CANUM.

NEBULA H V. 47 URSÆ MAJORIS.

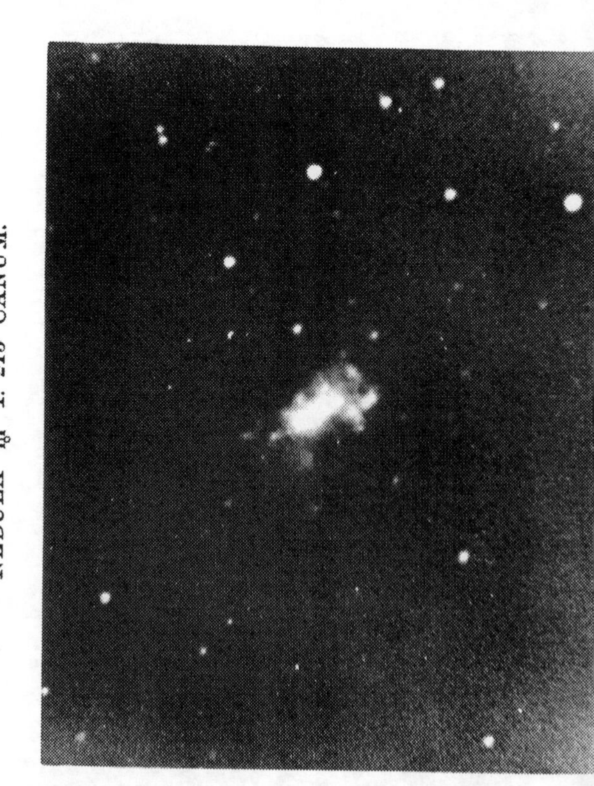

NEBULA H I. 213 CANUM.

PLATE XVII.

Spiral Nebula ♅ V. 47 Ursæ Majoris.

R.A. 9h. 55m. 9s. Dec. N. 56° 10′·1.

The photograph covers 20′·4 in R.A. and 16′·3 in Declination, and was taken with the 20-inch reflector on April 14th, 1895, between sidereal time 10h. 33m. and 12h. 3m., with an exposure of the plate during ninety minutes.

Scale—1 millimètre to 12 seconds of arc.

REFERENCES.

N.G.C. 3079. G.C. 1983.

Sir J. Herschel, in the G.C., describes the nebula as very bright; large; very much extended 135° ±.

Lord Rosse (*Obs. of Neb. and Cl.*, p. 81) describes it as very bright, very much extended *p.f.*, very much brighter in the middle, many bright stars near.

The photograph shows the nebula to be a spiral viewed nearly edgewise and extended in the direction 165° ±, which, it will be observed, differs considerably from the angles given by Herschel and Rosse. There is no nucleus visible, but the nebulosity is dense about the centre. One bright star, four faint ones, and some stellar condensations are visible in the nebulosity; but the convolutions cannot be seen because they are presented edgewise to the line of sight.

The nebula ♅ III. 853 is also shown on the plate near the *s.p.* edge, and appears like a bright star surrounded by faint nebulosity; it has a small indentation on its *s.f.* edge, which is apparently a space between two small *comites*. There is also on the plate a nebula, not referred to in the catalogues, about 5′ 20″ of arc *preceding* the bright star near the *n.* end of ♅ V. 47. It has a stellar nucleus surrounded by faint nebulosity, extended in the *preceding* direction.

PLATE XVII.

Spiral Nebula H V. 46 Ursæ Majoris.

R.A. 11h. 5m. 40s. Dec. N. 56° 13′·0.

The photograph covers 20′·4 in R.A. and 16′·3 in Declination, and was taken with the 20-inch reflector on April 20th, 1895, between sidereal time 10h. 43m. and 14h. 46m., with an exposure of the plate during four hours.

Scale—1 millimètre to 12 seconds of arc.

REFERENCES.

N.G.C. 3556. G.C. 2318. *h* 831.

Sir J. Herschel, in the G.C., describes the nebula as considerably bright; very large; very much extended 79°; pretty bright in the middle; mottled.

Lord Rosse (*Obs. of Neb. and Cl.*, p. 92-3) describes it as a curiously twisted nebula; light mottled; very bright in the middle; knot in *p.* branch; three stars involved. Query, extended spiral? Six observations are recorded and some measurements given.

The photograph shows the nebula to be a spiral with a faint elongated nucleus, densest at the centre; there is one bright star *preceding* it which seems too well defined to be actually involved; it is probably between us and the nebula. There are also six small stars, together with several nebulous condensations involved in the nebula; there are also places near the *preceding* half with little, if any, nebulosity in them; but the *following* half is not free from nebulosity; these vacancies are probably divisions between the convolutions.

PLATE XVII.

Spiral Nebula H I. 213 Canum Venaticorum.

R.A. 12h. 23m. 21s. Dec. N. 44° 38′·7.

The photograph covers 20′·4 in R.A. and 16′·3 in Declination, and was taken with the 20-inch reflector on April 24th, 1898, between sidereal time 11h. 50m. and 13h. 20m., with an exposure of the plate during ninety minutes.

Scale—1 millimètre to 12 seconds of arc.

REFERENCES.

N.G.C. 4449. G.C. 3002. *h* 1281.

Sir J. Herschel, in the G.C., describes the nebula as very bright; considerably large; much extended 15°; resolvable; star 9th mag. 5′ distant.

Lord Rosse (*Obs. of Neb. and Cl.*, p. 116) states that the nebula has three nuclei (or two nuclei and one star), and faint nebulosity outlying.

The photograph shows the nebula to be irregular in outline, with dense nebulosity in the *n.f.* to *s.p.* direction, and one bright star, together with a smaller one, and two or three condensations of nebulosity involved in it. On the north side are five or six stars involved in nebulosity; and on the north, south, and *preceding* sides are four well developed nebulous stars, besides several nebulous condensations.

PLATE XVII.

Spiral Nebula H I. 197-8 Canum Venaticorum.

R.A. I. 197. 12h. 25m. 40s. Dec. N. 42° 15′·3.
,, I. 198. 12h. 25m. 47s. ,, N. 42° 12′·0.

The photograph covers 20′·4 in R.A. and 16′·3 in Declination, and was taken with the 20-inch reflector on April 23rd, 1898, between sidereal time 11h. 46m. and 13h. 46m., with an exposure of the plate during two hours.

Scale—1 millimètre to 12 seconds of arc.

REFERENCES.

No. 197 N.G.C. 4485. G.C. 3041. *h* 1306.
,, 198 ,, 4490. ,, 3042. *h* 1308.

Sir J. Herschel, in the G.C., describes 197 as bright; pretty small; irregularly round; preceding of two; and 198 as very bright; very large; much extended 130°; partly resolvable.

Lord Rosse (*Obs. of Neb. and Cl.*, p. 117) suspected spirality in 197; the large nebula (198) has a star in the *f.* extremity, and an appendage north of the nucleus, a little following it. A drawing is given in the *Phil. Trans.*, 1861, pl. XXVII., fig. 23.

The photograph shows the nebula 197 to be a small spiral with dense nebulosity in the centre, and two or three stars involved; there is faint nebulosity on the *s.p.* side, with three small stars and two or three condensations of nebulosity involved. The nebula 198 is a right hand spiral, with a dense stellar nucleus in the midst of nebulosity; two of the convolutions have several star-like condensations with nebulosity around them. Much nebulosity is shown on the *s.p.* side, and some also on the *n.f.* side, in which is a prominent star-like condensation surrounded by nebulosity. In other parts of the faint nebulosity are several small, well-defined, star-like condensations; and also a prominent bright star in very faint nebulosity at the *s.f.* extremity.

Plate 18.

NEBULA H V. 42 COMÆ.

NEBULA H V. 25 CETI.

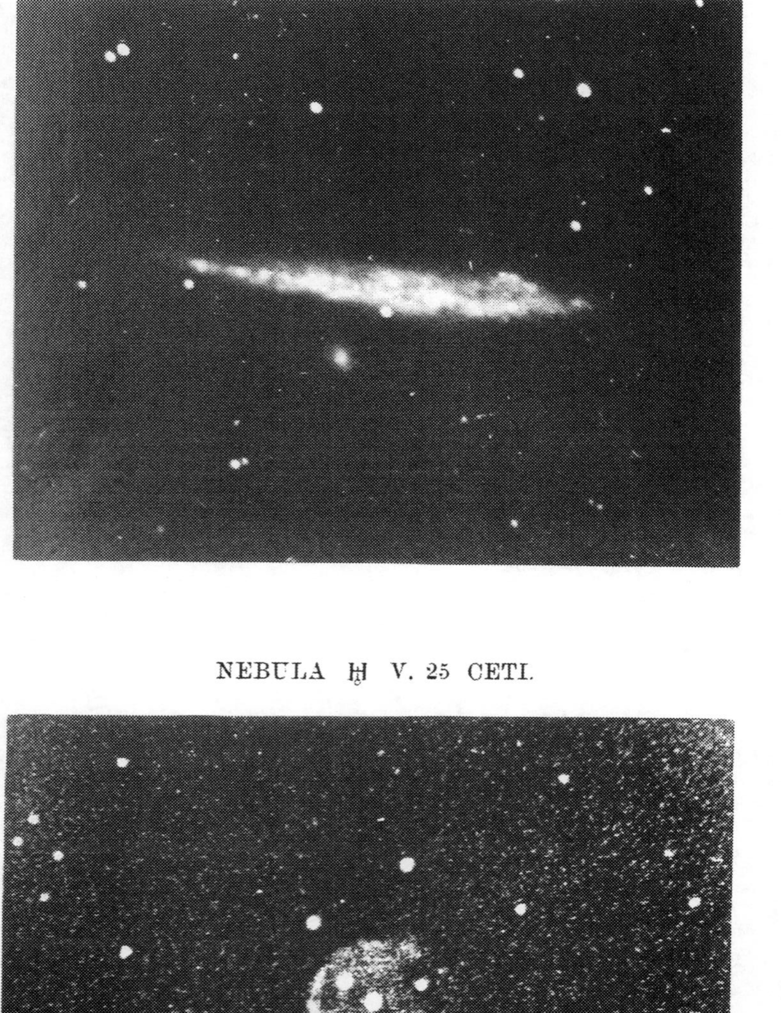

NEBULA H I. 176-7 COMÆ.

NEBULA H IV. 39 ARGÛS.

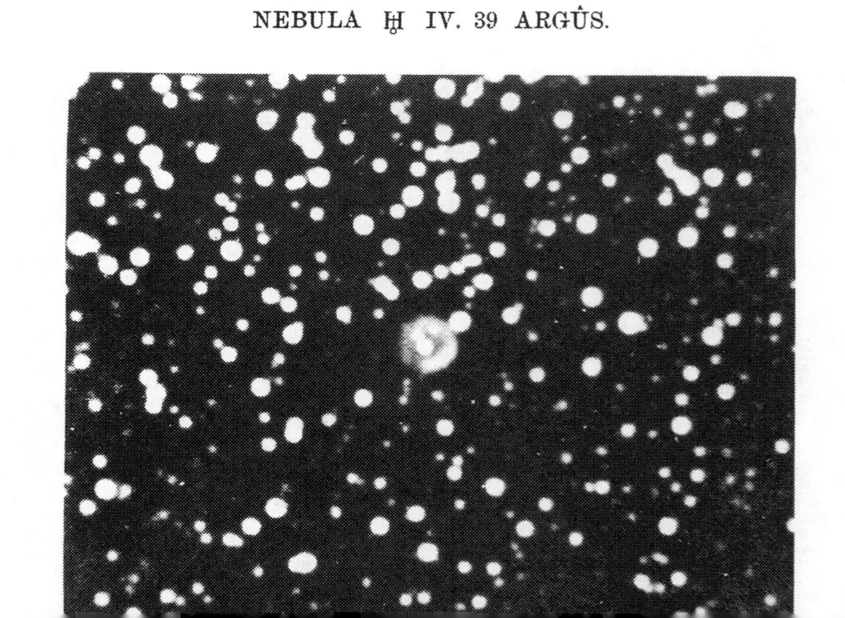

PLATE XVIII.

Spiral Nebula ꟸ. V. 42 Comæ Berenicis.

R.A. 12h. 37m. 19s. Dec. N. 33° 6′·0.

The photograph covers 20′·4 in R.A. and 16′·3 in Declination, and was taken with the 20-inch reflector on March 29th, 1894, between sidereal time 10h. 44m. and 13h. 47m., with an exposure of the plate during three hours.

Scale—1 millimètre to 12 seconds of arc.

REFERENCES.

N.G.C. 4631. G.C. 3165. *h* 1397.

Sir J. Herschel, in the G.C., describes the nebula as a remarkable object; very bright; very large; exceedingly extended, 70°±; brighter in the middle, with a nucleus; bright star near. A drawing is given in the *Phil. Trans.*, 1833, pl. XV., fig. 76.

Lord Rosse (*Obs. of Neb. and Cl.*, p. 119) states that it is a most extraordinary object, with a bright star near the centre, and at the right (north) masses of light appear through it in knots. A drawing is given in the *Phil. Trans.*, 1850, pl. XXXVII., fig. 9.

Lassell, in the *Mem. R.A.S.*, Vol. XXXVI., p. 46, gives a drawing, pl. V., fig. 24.

The photograph shows the nebula to be a spiral, seen nearly edgewise. It has a central stellar nucleus involved in nebulosity, with many stars, both bright and faint, together with nebulous condensations involved in the convolutions. The two bright stars—one near the centre, touching the *n.* edge, and the other near the *p.* extremity—do not appear to form part of the nebula, and are probably on the solar side of it. The nebula ꟸ II. 659 is north of the centre of the large nebula, and consists of a stellar centre, surrounded by nebulosity.

PLATE XVIII.

Spiral Nebulæ H I. 176-7 Comæ Berenicis.

R.A. I. 176. 12h. 39m. 6s. Dec. N. 32° 42′·8.
R.A. I. 177. 12h. 39m. 16s. Dec. N. 32° 46′·0.

The photograph covers 20′·4 in R.A. and 16′·3 in Declination, and was taken with the 20-inch reflector on March 29th, 1894, between sidereal time 10h. 44m. and 13h. 47m., with an exposure of the plate during three hours.

Scale—1 millimètre to 12 seconds of arc.

REFERENCES.

No. 176. N.G.C. 4656. G.C. 3189. *h* 1414.

No. 177. N.G.C. 4657. G.C. 3190. *h* 1415.

Sir J. Herschel, in the G.C., describes 176 as a remarkable object; pretty bright; large; very much extended 34°·3; *s.p.* of 2. Also 177 as a remarkable object; pretty faint; large; extended 90°±. A drawing is given in the *Phil. Trans.*, 1833, pl. XV., fig. 75.

Lord Rosse (*Obs. of Neb. and Cl.*, p. 120) states that Herschel's two nebulæ form one; the joining part in the middle faint; and faint prolongation of nebulosity as shown in the sketch in the *Phil. Trans.*, 1861, pl. XXVIII., fig. 26. Thinks he saw two additional stars in the *s.f.* extremity.

The photograph shows that the two nebulæ form one right-hand spiral, with a bright stellar nucleus involved in dense nebulosity; the *n.f.* convolution is much brighter than the others on the *s.f.* side, and there are several bright stars, and star-like condensations, involved in the bright convolutions; and also several faint condensations in the *s.p.* part.

The suspected two stars seen by Rosse at the *n.f.* extremity form a part of the bright convolution. On the *n.p.* side of the bend in the convolution is a small nebulous star, detached from the nebula, but which probably forms a part of it.

PLATE XVIII.

Spiral Nebula H V. 25 Ceti.

R.A. 0h. 42m. 2s. Dec. S. 12° 25′·3.

The photograph covers 20′·4 in R.A. and 16′·3 in Declination, and was taken with the 20-inch reflector on November 10th, 1896, between sidereal time 0h. 33m. and 3h. 28m., with an exposure of the plate during two hours and fifty-five minutes.

Scale—1 millimètre to 12 seconds of arc.

REFERENCES.

N.G.C. 246. G.C. 131. *h* 56.

Sir J. Herschel, in the G.C., describes the nebula as very faint; large; four stars in diffused nebulosity.

The photograph shows the nebula to be a very faint spiral, with three bright stars apparently involved, and one nearly touching the *n.p.* edge. I think these stars, like others already referred to, are not really involved in the nebula but are seen in projection upon it. There are faint condensations of nebulosity, those near the margin being the most prominent.

PLATE XVIII.

Annular Nebula ℌ IV. 39 Argûs.

R.A. 7h. 37m. 16s. Dec. S. 14° 30′·4.

The photograph covers 13′·6 in R.A. and 10′·8 in Declination, and was taken with the 20-inch reflector on February 24th, 1894, between sidereal time 6h. 2m. and 7h. 32m., with an exposure of the plate during ninety minutes.

Scale—1 millimètre to 8 seconds of arc.

REFERENCES.

N.G.C. 2438. G.C. 1565. *h* 464.

Sir J. Herschel, in the G.C., describes it as a planetary nebula; involved in the cluster; pretty bright; pretty small; extremely little extended; barely resolvable.

Lord Rosse (*Phil. Trans.*, 1850, p. 513, pl. XXXVIII., fig. 12, and *Obs. of Neb. and Cl.*, p. 61) records twenty-one observations of the nebula, and recognised it as an annular nebula, with two stars in it and a suspected third star.

Lassell, in the *Mem. R.A.S.*, Vol. XXIII., p. 60, pl. II., fig. 5, describes it as a planetary nebula, with two stars in it, and is quite separate from the cluster.

The photograph shows the nebula to be annular in form, strongly resembling *M.* 57 *Lyræ;* the annulus is most condensed on the *n.f.* side and there are three stars in the interior; the brightest of them (13-14 mag.) is near the centre; another of about 16th mag. is on the *s.p.* side; and the third, which is below 16th mag., is almost involved in the ring on the *s.f.* side. There is also evidence of very faint condensations of nebulosity in the ring itself.

Plate 19.

NEBULA M. 97 URSÆ MAJORIS.

NEBULA H I. 143 VIRGINIS.

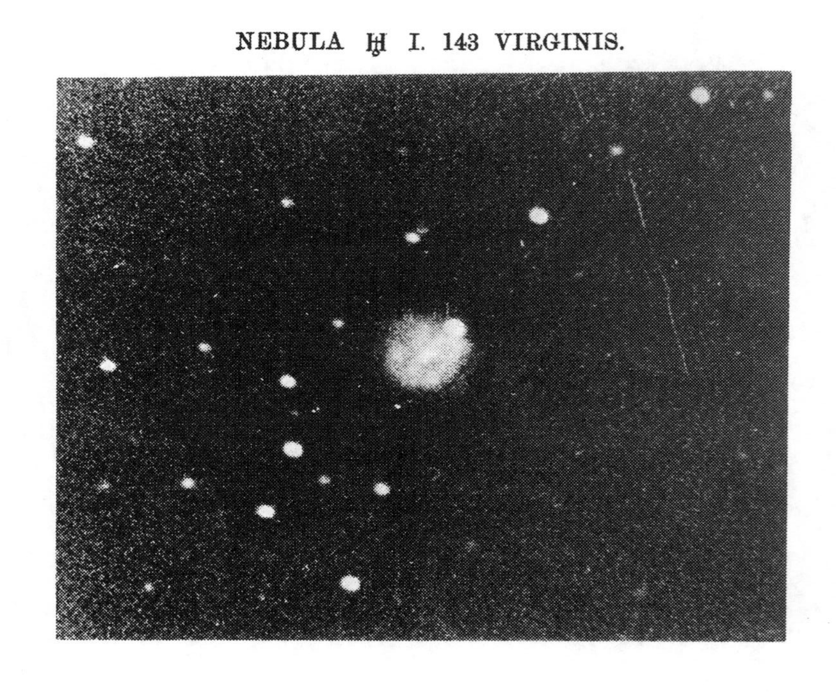

NEBULA M. 57 LYRÆ.

NEBULA H IV. 13 CYGNI.

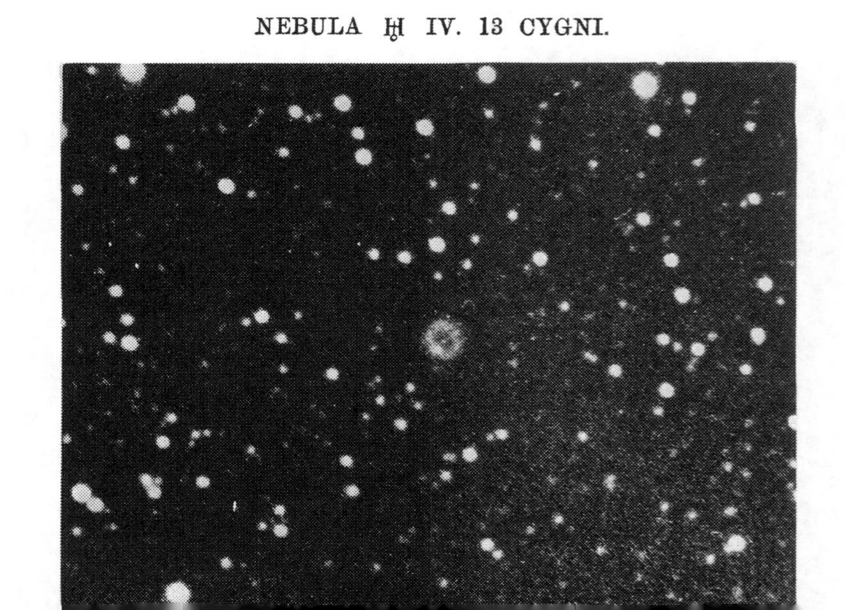

PLATE XIX.

Annular Nebula M. 97 Ursæ Majoris.

R.A. 11h. 9m. 0s. Dec. N. 55° 33′·7.

The photograph covers 20′·4 in R.A. and 16′·3 in Declination, and was taken with the 20-inch reflector on April 20th, 1895, between sidereal time 10h. 43m. and 14h. 46m., with an exposure of the plate during four hours.

Scale—1 millimètre to 12 seconds of arc.

REFERENCES.

N.G.C. 3587. G.C. 2343. *h* 838.

Sir J. Herschel, in the G.C., describes the nebula as a very remarkable object; a planetary nebula; very bright; very large; round; 19·0s. in diameter. It is figured in the *Phil. Trans.*, 1833, pl. X., fig. 25.

Lord Rosse (*Obs. of Neb. and Cl.*, p. 93, and *Phil. Trans.*, 1850, pl. XXXVII., fig. 11) figured the nebula as a circle inclosing two stars, and with hairy appendages round it, resembling in the whole the face of an Owl; and since then it has been known as the *Owl nebula.* Forty-five observations of the object were made between the years 1848 and 1874, in some of which both Lord Rosse and Dr. Robinson saw a faint star to the right of the central star, and suspected the existence of one or two other very faint stars, as well as a spiral shape; but he does not confirm the hair-like surroundings of the nebula.

The photograph shows the nebula as an ellipse (doubtless owing to perspective) with the major axis in *n.f.* to *s.p.* direction. It measures about 203 seconds of arc in length, and the star referred to by *Rosse* is very conspicuously seen in the centre, its magnitude being about the 15th, but there is no other star anywhere in the nebula, though there are two very faint condensations of nebulosity near the *s.p.* margin. The ring is not of equal breadth, but is widened on the *n.f.* and *s.p.* sides, and there is an absence of structure in the nebulosity; the photograph does not indicate any nebulous projections beyond the symmetrical outline of the nebula, and the star seen by both Lord Rosse and Dr. Robinson has disappeared.

PLATE XIX.

Nebula H I. 143 Virginis.

R.A. 12h. 55m. 36s. Dec. N. 3° 2′·0.

The photograph covers 13′·6 in R.A. and 10′·8 in Declination, and was taken with the 20-inch reflector on April 9th, 1894, between sidereal time 9h. 55m. and 12h. 57m., with an exposure of the plate during three hours.

Scale—1 millimètre to 8 seconds of arc.

REFERENCES.

N.G.C. 4900. G.C. 3356. *h* 1509.

Sir J. Herschel, in the G.C., describes the nebula as considerably bright; considerably extended; star 10th mag., attached at about 135°. It is figured in the *Phil. Trans.*, 1833, pl. XIV. fig. 67.

Lord Rosse (*Obs. of Neb. and Cl.*, pp. 123, 124) states that he saw it sometimes like the *Owl Nebula* (*h* 838) with a bright extended patch in the centre, and dark spots on each side of it. Sometimes a dark ring was seen all the way round, but blackest on the *s.p.* and *n.f.* sides. A marginal sketch is given.

The photograph shows the nebula to somewhat resemble the letter D, with the curved side in the *n.p.* direction, and a star of about 15th mag. in the centre, the interior being filled with nebulosity of different densities, within which are five or six star-like condensations; but there is no spiral structure visible, though the nebulous condensations resemble those invariably seen in spiral nebulæ. The *s.f.* part of the nebula is almost a straight line, with a star of about 12th mag. close to it, if not in actual contact.

PLATE XIX.

Annular Nebula M. 57 Lyræ.

R.A. 18h. 49m. 52s. Dec. N. 32° 54′·3.

The photograph covers 13′·6 in R.A. and 10′·8 in Declination, and was taken with the 20-inch reflector on July 10th, 1898, between sidereal time 17h. 42m. and 18h. 2m., with an exposure of the plate during 20 minutes.

Scale—1 millimètre to 8 seconds of arc.

REFERENCES.

N.G.C. 6720. G.C. 4447. *h* 2023.

Sir J. Herschel, in the G.C., describes the nebula as a magnificent object; annular; bright; pretty large; considerably extended, and gives a drawing of it in the *Phil. Trans.*, 1833, pl. X., fig. 29.

Lord Rosse (*Obs. of Neb. and Cl.*, p. 152) records eleven observations out of twenty-two made between 1848 and 1875, and gives a drawing in the *Phil. Trans.*, 1844, pl. XIX., fig. 29. He states that the filaments proceeding from the edge become more conspicuous under increasing magnifying power, within certain limits, which is strikingly characteristic of a cluster. Still, he did not feel confident that it was resolvable.

A photograph and descriptive matter relating to the nebula were given in *I.R. Photos.*, p. 107. The photograph was taken on July 27th, 1891, and exposure of 30 minutes. The present plate is given for comparison with it after the lapse of an interval of seven years; there is no appearance of stars involved in the ring, but the *s.f.* and *n.p.* sides are denser than the *s.p.* and *n.f.*—as was also the case in the early photograph; there is also an extension of faint nebulosity beyond the margins at each end of the major axis, as was observed by Lord Rosse and the late Mr. Pratt.

PLATE XIX.

Annular Nebula H IV. 13 Cygni.

R.A. 20h. 12m. 22s. Dec. N. 30° 15′·4.

The photograph covers 13′·6 in R.A. and 10′·8 in Declination, and was taken with the 20-inch reflector on August 31st, 1897, between sidereal time 20h. 27m. and 21h. 27m., with an exposure of the plate during sixty minutes.

Scale—1 millimètre to 8 seconds of arc.

REFERENCES.

N.G.C. 6894. G.C. 4565. *h* 2072.

Sir J. Herschel, in the G.C., describes it as a remarkable object ; annular ; faint ; small ; very very little extended, and gives a drawing in the *Phil. Trans.*, 1833, pl. XIII., fig. 48.

Lord Rosse (*Obs. of Neb. and Cl.*, p. 156) describes it as a fine annular nebula, like that in *Lyra* ; round ; the dark spaces extended *p.* to *f.* ; star easily seen in *n.p.* edge, and others suspected (1851). In 1855 there was a conspicuous star on the inner edge at *n.p.* side, and another, more difficult, near the *n.f.* edge, and he believed that the whole corner of the annulus was mottled. A drawing is given on pl. V.

The photograph shows the nebula as a nearly circular ring, resembling *M.* 57 *Lyræ*, with a small faint star near the centre and a bright star on the *n.p.* side of the ring ; there are slight irregularities in the density of the nebulosity forming the ring.

Plate 20.

NEBULA H V. 41 CANUM.

NEBULA H I. 43 VIRGINIS.

NEBULA H V. 8 LEONIS.

NEBULA H V. 24 COMÆ.

PLATE XX.

Nebula H V. 8 Leonis.

R.A. 11h. 15m. 3s. Dec. N. 14° 8′·4.

The photograph covers 20′·4 in R.A. and 16′·3 in Declination, and was taken with the 20-inch reflector on February 28th, 1894, between sidereal time 8h. 19m. and 12h. 2m., with an exposure of the plate during three hours and forty minutes.

Scale—1 millimètre to 12 seconds of arc.

REFERENCES.

N.G.C. 3628. G.C. 2378. *h* 857.

Sir J. Herschel, in the G.C., describes the nebula as pretty bright; very large; very much extended in the direction 102°; a drawing is given in the *Phil. Trans.*, 1833, pl. XIV., fig. 51.

Lord Rosse (*Obs. of Neb. and Cl.*, p. 95, and pl. III., fig. 8) describes and delineates it as pretty bright; extended; split into two parallel rays; the split extending its whole length, the *following* part being partially filled with faint nebulosity.

The photograph shows it like a spiral viewed edgewise, with a large dense central condensation; the ring being divided along its periphery into two parts parallel with each other by a broad dark band, or ring, which shuts from view the light of the central condensation. The two extremities of the diameter of the supposed ring show expansions of the nebulosity; there are two stars apparently involved. I have but very little doubt that these appearances are caused by the fainter parts of the convolutions of the nebula being now turned towards the earth, and that they obscure the nucleus and the brighter parts of the nebula, but are not bright enough themselves to affect the photographic film, therefore they produce the dark line.

PLATE XX.

Spiral Nebula H V. 41 Canum Venaticorum.

R.A. 12h. 12m. 29s. Dec. N. 38° 22′·0.

The photograph covers 20′·4 in R.A. and 16′·3 in Declination, and was taken with the 20-inch reflector on April 28th, 1897, between sidereal time 11h. 30m. and 13h. 0m., with an exposure of the plate during ninety minutes.

Scale—1 millimètre to 12 seconds of arc.

REFERENCES.

N.G.C. 4244. G.C. 2831. *h* 1167.

Sir J. Herschel, in the G.C., describes the nebula as pretty bright; very large; extremely extended 43°·2; very gradually brighter in the middle.

Lord Rosse (*Obs. of Neb. and Cl.*, p. 112) describes it as a very bright lenticular ray *s.p.* to *n.f.* about 18′ long; very gradually brighter in the middle, almost to a nucleus. He saw a star in the centre several times; a double star in the *p.* edge *n.* of the centre, and in the *s.p.* end what might be called a very faint triple star; faint star involved near the *n.f.* end.

The photograph shows the nebula to be a spiral, seen edgewise, with a stellar nucleus in the midst of nebulosity. There are about twelve star-like condensations involved in the nebulosity, which extends in *n.f.* to *s.p.* direction about 13·5 minutes of arc; and this represents the length of the nebula, its breadth being very narrow.

The stars in this region of the sky are small and comparatively few in number.

PLATE XX.

Spiral Nebula H V. 24 Comæ Berenicis.

R.A. 12h. 31m. 24s. Dec. N. 26° 32′·2.

The photograph covers 20′·4 in R.A. and 16′·3 in Declination, and was taken with the 20-inch reflector on May 11th, 1896, between sidereal time 12h. 49m. and 15h. 42m., with an exposure of the plate during two hours and fifty-three minutes.

Scale—1 millimètre to 12 seconds of arc.

REFERENCES.

N.G.C. 4565. G.C. 3106. *h* 1357.

Sir J. Herschel, in the G.C., describes the nebula as bright; extremely large; extremely extended 136°·1; very suddenly brighter in the middle; stars 10-11 mag. A drawing is given in the *Phil. Trans.*, 1833, pl. XII., fig. 37.

Lord Rosse (*Obs. of Neb. and Cl.*, p. 118) describes the nebula as a ray, and gives a marginal sketch of it. The ray is 12′ or 15′ long and there is a faint star in it; extended about 130°.

The photograph shows the nebula to be, almost certainly, a spiral viewed edgewise, the dark line across it being caused by the fainter portion of the nebulous convolutions being now turned towards the earth; they would thus be dense enough to obscure the nucleus and its surroundings but not bright enough themselves to impress the film; they thus appear as a dark line. The nebula subtends 15 minutes of arc, which represents its length, and from this its breadth can be inferred.

PLATE XX.

Nebula H I. 43 Virginis.

R.A. 12h. 34m. 47s. Dec. S. 11° 4′·3.

The photograph covers 20′·4 in R.A. and 16′·3 in Declination, and was taken with the 20-inch reflector on April 27th, 1897, between sidereal time 13h. 6m. and 14h. 36m., with an exposure of the plate during ninety minutes.

Scale—1 millimètre to 12 seconds of arc.

REFERENCES.

N.G.C. 4594. G.C. 3132. *h* 1376.

Sir J. Herschel, in the G.C., describes it as a remarkable object ; very bright ; very large ; extremely extended 92° ; very suddenly much brighter in the middle ; with a nucleus. A drawing of it is given in the *Phil. Trans.*, 1833, pl. XIV., fig. 50.

Lassell, in the *Mem. R.A.S.*, Vol. XXXVI., p. 45, describes it as a very long spindle-shaped nebula ; its axis almost parallel with the equator ; it has a dark band through its entire length which extends to 7′ 5″ ; there are four stars in the field. A drawing of it is given on pl. V., fig. 22.

The photograph shows the nebula to be probably a spiral, and that the dark line across it is due to the less dense portion of the convolutions being now turned edgewise in the direction of the earth, thus obscuring a part of the nucleus and of its dense surroundings ; while the light of those parts of the convolutions is too feeble to affect the film. The line therefore appears dark on the photograph, and indicates the thickness or depth of the nebula.

Plate 21.

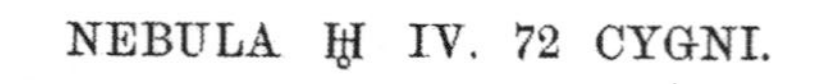

NEBULA H IV. 72 CYGNI.

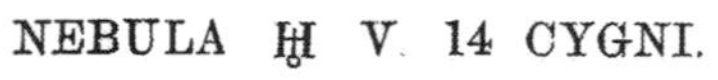

NEBULA H V. 14 CYGNI.

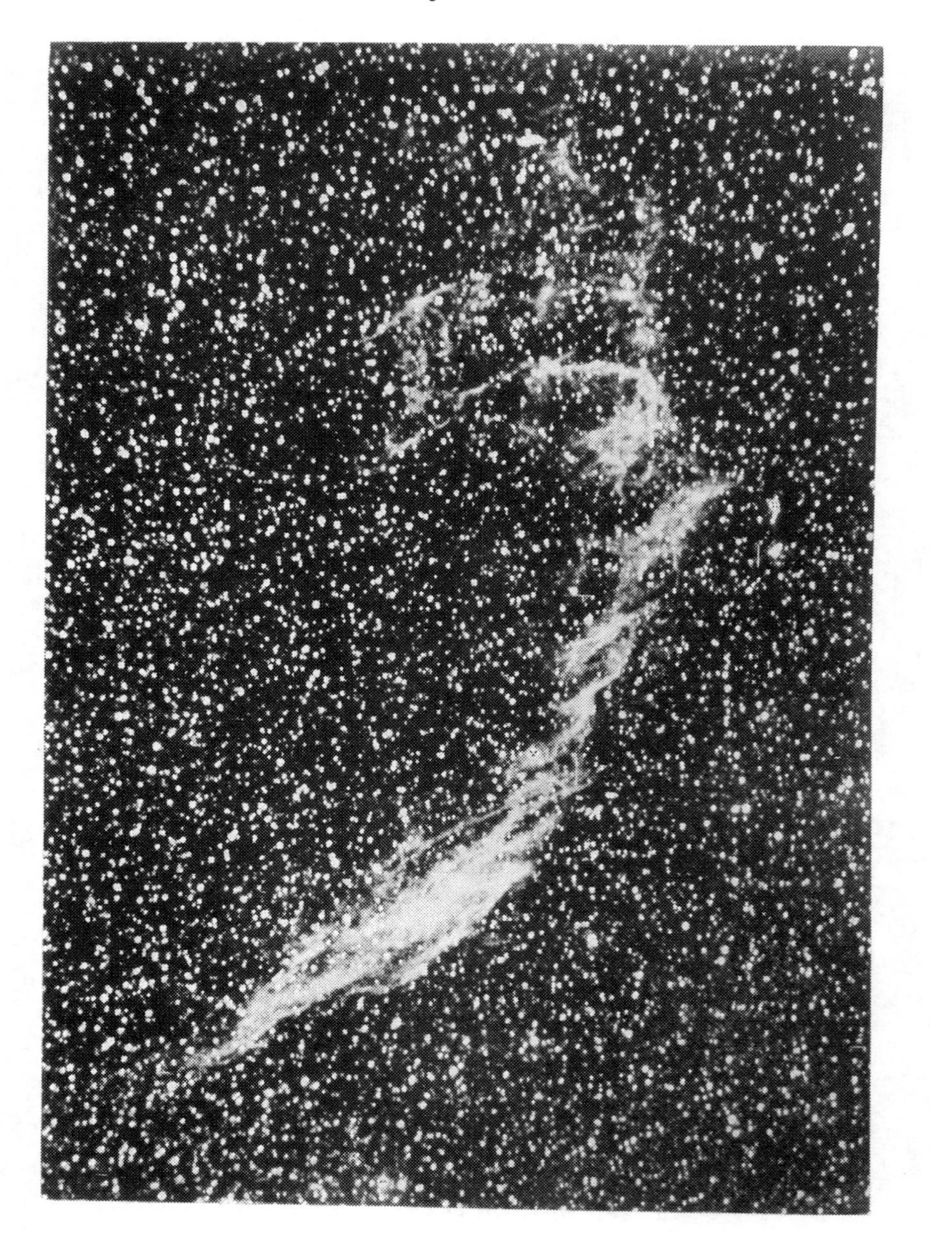

PLATE XXI.

Nebula H IV. 72 Cygni.

R.A. 20h. 8m. 50s. Dec. N. 38° 5′·5.

The photograph covers the region between R.A. 20h. 6m. 11s. and R.A. 20h. 11m. 5s. Declination between 37° 26′·9 and 38° 49′·4 North.

Scale—1 millimètre to 30 seconds of arc.

Co-ordinates of the Fiducial stars marked with dots for the epoch A.D. 1900.

Star (.)	D.M. No. 3814—Zone + 37°	... R.A. 20h. 7m. 48·1s.	... Dec. N. 37° 45′·2	... Mag. 9·1
„ (··)	„ „ 3821 „ 37°	... „ 20h. 8m. 28·5s.	... „ 38° 3′·3	... „ 7·1
„ (∵)	„ „ 3956 „ 38°	... „ 20h. 9m. 46·3s.	... „ 38° 27′·8	... „ 7·2

The photograph was taken with the 20-inch reflector on September 3rd, 1897, between sidereal time 20h. 7m. and 22h. 58m., with an exposure of the plate during two hours and fifty-one minutes.

REFERENCES.

N.G.C. 6888. G.C. 4561.

Sir J. Herschel, in the G.C., describes the nebula as very faint; large; little extended.

Lord Rosse (*Obs. of Neb. and Cl.*, p. 155 and pl. V.) states that the nebula is very, very large; very faint; very much extended; fades gradually in all directions; milky to a considerable distance.

The photograph shows the nebula to be elliptic in outline, and the area within to be filled with flocculent nebulosity. A large number of normal stars are studded over the whole area; but I do not think that they are in the same plane as the nebula. There is also visible a large number of small nebulous stars and of star-like condensations of nebulosity involved in it. The *north preceding* margin of the nebula is bright and has a large number of faint nebulous stars involved; but it is possible that they, or some of them, may be in space beyond, and are seen because their light shines through the nebula.

PLATE XXI.

Nebula H V. 14 Cygni.

R.A. 20h. 52m. 14s. Dec. N. 31° 18′·8.

The photograph covers the region between R.A. 20h. 49m. 59s. and 20h. 54m. 26s. Declination between 30° 21′·9 and 31° 44′·4 North.

Scale—1 millimètre to 30 seconds of arc.

Co-ordinates of the Fiducial stars marked with dots for the epoch A.D. 1900.

Star	D.M. No.	Zone	R.A.	Dec. N.	Mag.
Star (.)	D.M. No. 4229	—Zone +30° ...	R.A. 20h. 50m. 2·9s. ...	Dec. N. 30° 41′·3 ...	Mag. 9·1
„ (··)	„ „ 4268	„ 31° ...	„ 20h. 51m. 46·8s. ...	„ 31° 42′·8 ...	„ 7·6
„ (∴)	„ „ 4279	„ 31° ...	„ 20h. 52m. 35·8s. ...	„ 31° 12′·4 ...	„ 8·5
„ (::)	„ „ 4254	„ 30° ...	„ 20h. 53m. 54·3s. ...	„ 30° 55′·1 ...	„ 8·0

The photograph was taken with the 20-inch reflector on November 4th, 1896, between sidereal time 21h. 18m. and 0h. 13m., with an exposure of the plate during two hours and fifty-five minutes.

REFERENCES.

N.G.C. 6992. G.C. 4616. *h* 2092.

Sir J. Herschel and Lord Rosse have observed and described this nebula, but it is obvious that no description, or hand delineation, could possibly convey an accurate idea of this complex object.

The photograph shows the nebula to be of a wave-like form, extending a distance of 80 minutes of arc in *south following* to *north preceding* direction, and that there is evidence of disturbing influences in several parts. On the southern part is a cylindrical aggregation of nebulosity, seen more clearly on the negative, and the whole area of the nebula as well as the region surrounding it are densely studded with normal stars of between the 10th and 17th magnitude. A photograph, measuring 17 degrees in diameter, was simultaneously taken with the 5-inch Cooke lens, and it shows that there are patches of wave-like nebulosity extending over an area of two and a half degrees in the *preceding* direction from this, and includes the nebula H V. 15 *Cygni*. N.G.C. 6960.

NEBULA N.G.C. 1499 PERSEI.

NEBULA N.G.C. 281 CASSIOPEIÆ.

PLATE XXII.

Nebula N.G.C. 281 Cassiopeiæ

R.A. 0h. 47m. 26s. Dec. N. 56° 3′ 0.

The photograph covers the region between R.A. 0h. 43m. 45s. and 0h. 50m. 30s. Declination between 55° 24′·3 and 56° 46′·8 North.

Scale—1 millimètre to 30 seconds of arc.

Co-ordinates of the Fiducial stars marked with dots for the epoch A.D. 1900.

Star (.) D.M. No. 179—Zone + 55°	...	R.A.	0h. 45m. 40·5s.	...	Dec. N. 56° 7′·0	...	Mag. 9·2
„ (··) „ „ 143 „ 56°	...	„	0h. 47m. 13·2s.	...	„ 56° 40′·6	...	„ 7·4
„ (∵) „ „ 195 „ 55°	...	„	0h. 48m. 37·5s.	...	„ 55° 46′·6	...	„ 7·2

The photograph was taken with the 20-inch reflector on November 6th, 1896, between sidereal time 0h. 12m. and 3h. 6m., with an exposure of the plate during two hours and fifty-four minutes.

REFERENCES.

N.G.C. 281; where it is described as faint; very large; diffuse; a small triple star on *north preceding* edge.

The photograph shows the nebula to cover an area of about 27′ from *s.f.* to *n.p.* and about 20′ from *f.* to *p.*, with a large bay-like indentation on the *s.p.* side. The sky area of several square degrees is crowded with stars both bright and faint, and the nebulosity also is similarly studded with stars, but there are not many prominent and distinct stellar condensations involved in the nebulosity.

The black spots seen in the nebula are due to defects in the film.

PLATE XXII.

Nebula N.G.C. 1499 Persei.

R.A. 3h. 56m. 55s. Dec. N. 36° 8′·5.

The photograph covers the region between R.A. 3h. 49m. 21s. and 3h. 58m. 24s. Declination between 35° 41′·9 and 36° 58′·6 North.

Scale—1 millimètre to 40 seconds of arc.

Co-ordinates of the Fiducial stars marked with dots for the epoch A.D. 1900.

Star (.) D.M. No. 792—Zone + 36°	... R.A. 3h. 51m. 3·6s.	... Dec. N. 36° 11′·9	... Mag. 7·0
„ (··) „ „ 805 „ 36°	... „ 3h. 54m. 39·2s.	... „ 36° 43′·6	... „ 7·5

The photograph was taken with the 20-inch reflector on December 18th, 1897, between sidereal time 2h. 44m. and 4h. 14m., with an exposure of the plate during ninety minutes.

REFERENCES.

N.G.C. 1499, where it is described as very faint, very large, extended north and south, diffuse.

Dr. Scheiner, in the *Astronomische Nachrichten*, No. 3157, gives a drawing of it.

The photograph shows the nebula to be cloud-like in form and extended more than 2 degrees in *south following* to *north preceding* direction; in breadth it is 33′; much structural detail is visible, and its area is studded with many stars, and some star-like condensations in the aggregations of the nebulosity. The stars are in several curves and lines, but they do not indicate in a striking manner that they are physically connected with the nebula; they are probably between the earth and the nebula.

Plate 23.

NEBULA M. 16 CLYPEI.

NEW NEBULA IN CYGNUS.

PLATE XXIII.

Nebula M. 16 Clypei.

R.A. 18h. 13m. 13s. Dec. S. 13° 50′·0.

The photograph covers the region between R.A. 18h. 11m. 12s. and R.A. 18h. 15m. 11s. Declination between 13° 10′·0 and 14° 32′·5 South.

Scale—1 millimètre to 30 seconds of arc.

Co-ordinates of the Fiducial stars marked with dots for the epoch A.D. 1900.

Star	D.M. No.	Zone	R.A.	Dec. S.	Mag.
Star (.)	D.M. No. 4975	—Zone − 14° ...	R.A. 18h. 12m. 8·3s. ...	Dec. S. 14° 6′·8 ...	Mag. 9·0
„ (··)	„ „ 4916	„ 13° ...	„ 18h. 12m. 26·9s. ...	„ 13° 11′·7 ...	„ 8·3
„ (·:·)	„ „ 5005	„ 14° ...	„ 18h. 14m. 53·2s. ...	„ 14° 0′·0 ...	„ 8·9

The photograph was taken with the 20-inch reflector on August 4th, 1897, between sidereal time 18h. 50m. and 20h. 50m., with an exposure of the plate during two hours.

REFERENCES.

N.G.C. 6611. G.C. 4400. *h* 2006.

Sir J. Herschel, in the G.C., describes it as a cluster; at least 100 stars, large and small, in it.

The photograph shows that it is a large bright *nebula* with a cluster apparently involved in it. The nebula measures 23′ from *n.f.* to *s.p.*, and 17′·5 from *s.f.* to *n.p.* It is irregular in outline, with many stars and star-like condensations, and the cluster, together with many stars, are either involved in or are in alignment with the nebula. It is not recorded in the catalogues, and therefore is probably new.

PLATE XXIII.

Nebula in Cygnus.

R.A. 20h. 52m. 35s. Dec. N. 47° 2′·0.

The photograph covers the region between R.A. 20h. 49m. 32s. and R.A. 20h. 55m. 6s. Declination between 46° 21′·7 and 47° 44′·2 North.

Scale—1 millimètre to 30 seconds of arc.

Co-ordinates of the Fiducial stars marked with dots for the epoch A.D. 1900.

Star		D.M. No.	Zone		R.A.		Dec. N.		Mag.
Star	(.)	D.M. No. 3220	—Zone +47°	...	R.A. 20h. 51m. 17·6s.	...	Dec. N. 47° 22′·3	...	Mag. 8·8
„	(··)	„ „ 3111	„ 46°	...	„ 20h. 52m. 27·0s.	...	„ 47° 2′·0	...	„ 6·0
„	(·.·)	„ „ 3112	„ 46°	...	„ 20h. 52m. 29·4s.	...	„ 46° 42′·7	...	„ 7·8
„	(::)	„ „ 3240	„ 47°	...	„ 20h. 54m. 52·0s.	...	„ 47° 13′·5	...	„ 7·5

The photograph was taken with the 20-inch reflector on September 13th, 1895, between sidereal time 20h. 0m. and 22h. 0m., with an exposure of the plate during two hours.

REFERENCES.

Sir J. Herschel, in the G.C., describes it as a cluster; large; poor; very little compressed.

The photograph shows that there is no definite cluster, but that there is a *nebula* in which is partly involved the star D.M. 3111 Zone 46° R.A. 20h. 52m. 27·2s. Dec. N. 47° 2′·0 (marked (··) on the plate). The nebula is about 6′·5 in length from *n.* to *s.*, and 5′·0 from *p.* to *f.*; is irregular in outline, with many stars both bright and faint in alignment with it, and some star-like condensations in the nebulosity; its structure is of a fleecy character. The nebula is not recorded in the catalogues, and is therefore probably new.

Plate 24.

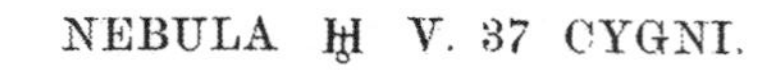

NEBULA ♅ V. 37 CYGNI.

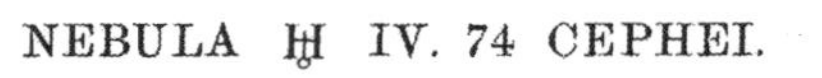

NEBULA ♅ IV. 74 CEPHEI.

PLATE XXIV.

Nebula H V. 37 Cygni.

R.A. 20h. 55m. 14s. Dec. N. 43° 56′·1.

The photograph covers the region between R.A. 20h. 51m. 36s. and R.A. 21h. 0m. 15s. Declination between 42° 58′·4 and 44° 55′·2 North.

Scale—1 millimètre to 48 seconds of arc.

Co-ordinates of the Fiducial stars marked with dots for the epoch A.D. 1900.

Star					R.A.	Dec.	Mag.
Star (·)	D.M. No.	3766	—Zone	+43°	R.A. 20h. 52m. 26·7s.	Dec. N. 43° 59′·4	Mag. 6·7
„ (··)	„	3911	„	42°	„ 20h. 52m. 34·7s.	„ 43° 2′·4	„ 6·7
„ (·:·)	„	3789	„	43°	„ 20h. 58m. 12·9s.	„ 43° 47′·4	„ 7·0

The photograph was taken with the 20-inch reflector on October 10th, 1896, between sidereal time 20h. 35m. and 23h. 30m., with an exposure of the plate during two hours and fifty-five minutes.

REFERENCES.

N.G.C. 7000. G.C. 4621. *h* 2096.

The photograph covers an area of about three square degrees in the sky and the nebula covers the larger portion of it; the extensions of the nebulosity reach a little beyond the limits of the photograph. There is no indication of a symmetrical arrangement of the nebulosity, though, in parts, there are disturbed areas with some concentration or aggregation of nebulous matter; and it will be observed that the whole area of the nebula is densely studded with stars, but as they do not follow the curvatures they are not probably really involved in the nebulosity; it is more probable that they lie between the earth and the nebula.

Such nebulous areas as this would very probably be crossed by other moving bodies in space and cause outbursts of light of short duration such as have been recorded from time to time.

PLATE XXIV.

Nebula H IV. 74 Cephei.

R.A. 21h. 0m. 30s. Dec. N. 67° 46′·2.

The photograph covers the region between R.A. 20h. 55m. 20s. and R.A. 21h. 5m. 20s. Declination between 67° 3′·0 and 68° 25′·5 North.

Scale—1 millimètre to 30 seconds of arc.

Co-ordinates of the Fiducial stars marked with dots for the epoch A.D. 1900.

Star	D.M. No.	Zone		R.A.		Dec. N.		Mag.
Star (.)	D.M. No. 1279—	Zone + 67°	...	R.A. 20h. 56m. 33s.	...	Dec. N. 67° 22′·4	...	Mag. 7·2
„ (··)	„ „ 1173	„ 68°	...	„ 20h. 58m. 40s.	...	„ 68° 24′·9	...	„ 8·8
„ (∵)	„ „ 1285	„ 67°	...	„ 21h. 3m. 39s.	...	„ 67° 35′·3	...	„ 8·8

The photograph was taken with the 20-inch reflector on September 18th, 1898, between sidereal time 21h. 1m. and 23h. 55m., with an exposure of the plate during two hours and fifty-four minutes.

REFERENCES.

N.G.C. 7023. G.C. 4634.

Sir J. Herschel, in the G.C., describes the nebula as extremely faint; query (?) star of 7th mag. in it.

The photograph shows the nebula with a bright stellar nucleus in the midst of dense nebulosity, which is of a cloud-like character, with several very faint stars apparently involved. The dark areas in it are rather remarkable, and give well defined boundaries to parts of the nebulosity.

The nebula appears in a region almost void of stars.

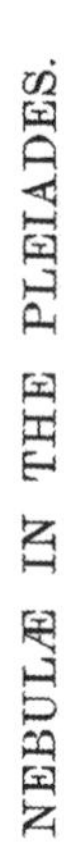

NEBULÆ IN THE PLEIADES.

NEBULÆ NEAR γ CASSIOPELÆ.

PLATE XXV.

The Region of Gamma Cassiopeiæ.

R.A. 0h. 50m. 40s. Dec. N. 60° 10′·6.

The photograph covers the region between R.A. 0h. 46m. 11s. and 0h. 55m. 10s. Declination between 59° 25′·1 and 61° 1′·4 North.

Scale—1 millimètre to 35 seconds of arc.

Co-ordinates of the Fiducial stars marked with dots for the epoch A.D. 1900.

Star	D.M. No.	Zone	R.A.	Dec. N.	Mag.
Star (.)	D.M. No. 124	—Zone + 60° ...	R.A. 0h. 47m. 7·5s. ...	Dec. N. 60° 33′·9 ...	Mag. 5·0
„ (··)	„ „ 146	„ 59° ...	„ 0h. 50m 45·6s. ...	„ 59° 49′·3 ...	„ 6·3
„ (∴)	„ „ 137	„ 60° ...	„ 0h. 51m. 16·1s. ...	„ 60° 53′·1 ...	„ 7·0
„ (::)	„ „ 161	„ 59° ...	„ 0h. 53m. 57·5s. ...	„ 59° 58′·3 ...	„ 7·2

The photograph was taken with the 20-inch reflector on October 25th, 1895, between sidereal time 0h. 16m. and 1h. 46m., with an exposure of the plate during ninety minutes.

REFERENCES.

Index Catalogue, No. 59-63. *Astron. Nachrichten*, No. 3214.

The star γ is involved in the centre of the patch of atmospheric glare caused by its light, but on the negative the reversed image of the star is seen as a small white spot, which marks a well defined point for the purposes of measurement. Two conical, or fan-shaped, nebulæ are shown on the *north following* side, and there is some very faint nebulosity, extending in semi-circular form, between them. Many stars are studded over the surfaces of the nebulæ, but without affording conclusive evidence that they are physically connected with them; there are also some stars that appear to be actually involved in the nebulosity. At the apex of each nebula is a faint star, which forms a convenient point for measuring positions and distances. Of these I give the following particulars:—

Pos. angle from γ to the star at the apex of the northernmost nebula (1) in the year 1895	14° 51′ 17″	Distance 1360″
Pos. angle of the same star in 1890	15° 1′ 53″	„ 1356″
Pos. angle of the star at the apex of the second nebula (2) in 1895 ...	57° 55′ 33″	„ 1158″
Pos. angle of the same star in 1890	57° 50′ 51″	„ 1158′

The two photographs here referred to were taken, one on the 17th January, 1890, and the other on the 25th October, 1895, the interval between them being 5 years and 281 days, and we find, as the result of the measurements of the two nebulæ, that a longer interval of time must elapse, and additional photographs taken and measured, before we shall be justified in forming positive conclusions concerning the movements of these nebulæ.

PLATE XXV.

Nebulæ in the Pleiades.

The photograph covers the region between R.A. 3h. 37m. 30s. and R.A. 3h. 44m. 16s. Declination between 22° 47′·6 and 24° 50′·0 North.

Scale—1 millimètre to 48 seconds of arc.

Co-ordinates of the Fiducial stars marked with dots for the epoch A.D. 1900.

Star (.)	D.M. No. 505—Zone + 23°	...	R.A. 3h. 38m. 52·1s.	...	Dec. N. 23° 58′·8	... Mag. 6·5
„ (··)	„ „ 562 „ 24°	...	„ 3h. 41m. 1·8s.	...	„ 24° 12′·8	... „ 7·5
„ (∵)	„ „ 537 „ 23°	...	„ 3h. 41m. 24·9s.	...	„ 23° 29′·5	... „ 7·5
„ (::)	„ „ 561 „ 23°	...	„ 3h. 43m. 24·7s.	...	„ 24° 4′·6	... „ 7·5

The photograph was taken with the 20-inch reflector on three nights as follows:—

December 22nd, 1897	...	Between sidereal time 23h. 55m. and 3h. 21m.	...	Exp.	3h. 20m.
„ 23rd „	...	„ „ 0h. 40m. „ 3h. 44m.	...	„	3h. 0m.
„ 25th „	...	„ „ 0h. 2m. „ 3h. 47m.	...	„	3h. 40m.

The total exposure being ten hours in duration, extending over an interval of four days.

A photograph of the region of the Pleiades was taken with the 20-inch reflector on December 8th, 1888, with an exposure of four hours, and is published in *I.R. Photos.*; and, when the two original negatives (one with four hours' and the other with ten hours' exposure, with an interval of nine years and seventeen days between them) are compared with each other, it is found that there is only a small difference in the extent of the nebulosity depicted upon them. The *Merope* nebula extends from the star to the distance of about 40 minutes of arc, and faintly covers an area of about 50 minutes in width, in the south, *south preceding*, and *south following* directions; the *Maia* nebulosity also extends about 40 minutes of arc in the *n.f.* direction. The other stars and nebulosities, which together form the Group of the *Pleiades*, show but little, if any, further extensions, or changes of form, during the interval of nine years and seventeen days, though the density of the photographic images is greater on the plates with the long exposures. The photograph annexed is one of a group of four that were taken of this region between December, 1893, and December, 1897, with respective exposures of eight to twelve hours, and they confirm the accuracy of each other in every particular.

In examining the original negative of the photograph now reproduced, and also the others referred to, the nebula round *Merope* is seen to be of a streaky character, with the star in the centre of the denser part; the streaks trend in *south following* to *north preceding* directions for a distance of about 16 minutes of arc, and then curve in the direction *south preceding*, where the nebulosity becomes very faint but still streaky in character, and spreads out to the extent of 50 minutes of arc in breadth. It does not seem to commingle in any degree with the other adjacent nebulosities, and there are four faint stars involved in it, within a distance of 60 seconds of arc from *Merope*. Besides these, there are other bright and faint stars involved in the nebulosity.

A straight streak of dense nebulosity extends across the two stars 1 and 7 (*Bessel*): it is about 19 minutes of arc in length, and does not seem to form a part of any of the other nebulosities. I infer that it is an independent spiral nebula, seen edgewise, and there is a bright star at its centre, which is probably the nucleus, and three faint stars apparently involved.

Electra does not seem to be involved in nebulosity, but there is a streak of strong nebulosity extending to a distance of about 14 minutes of arc from it towards the *following* direction; this streak is divided along its full length into two equal parts by a faint dark line. Six stars of between 13th and 16th magnitudes are, apparently, involved in this nebula, which is seen edgewise, and is, I think, independent of the others. The nebulosity which extends from *Maia* crosses the streak above referred to at an angle of 45 degrees, but no physical result seems to have been caused by this, and this fact confirms the idea of the separate existence of both nebulæ. *Alcyone* is surrounded by nebulosity of both a streaky and flocculent character. The streaks on the *following* side trend *south following* to *north preceding*, and those on the *preceding* side cross the streaks of the *Maia* nebula at nearly right angles, without appearance of either disturbance or of commingling. This again points to the nebulæ being independent. To the north of *Alcyone* is a streak of nebulosity 30 minutes of arc in length, slightly curved towards the north; it involves the stars 10 and 24 (*Bessel*) besides twelve other stars of between 12th and 18th magnitudes. This appears to be an independent nebula seen edgewise, and is probably of a spiral character. The north part of the *Alcyone* nebula seems to cross this streak without disturbance being caused in either of them.

Maia is surrounded by streaky nebulosity trending north and south in parts, and *north following* to *south preceding* in other parts. To the *north following* it extends in a long streaky arm a little beyond the star 12 (*Bessel*)—a distance of about 44 minutes of arc, and on the *north preceding* side reaches *Taygeta*. There is also evidence of very faint nebulosity extending 30 minutes of arc on the *south following* side towards *Maia*.

No clear proof is given that *Atlas* and *Pleione* are involved in nebulosity.

It appears to me that the evidence, taken as a whole, points strongly to the probability that the *Pleiades* consist of a group of stars seen by us either behind or in front of a group of nebulæ, *Alcyone, Merope,* and *Maia* being involved in nebulosity. At present there is no indication of physical connection of these with each other or with the other nebulæ in the group. I may further add that it will be exceedingly difficult work to extend, materially, the knowledge we now possess concerning the group until movements and changes in the structure of the nebulosity can be detected amongst its component parts.

NEBULA M. 42 ORIONIS.

NEBULA M. 1 TAURI.

PLATE XXVI.

The Great Nebula M. 42 Orionis.

R.A. 5h. 30m. 22s. Dec. S. 5° 27′·5.

The photograph covers the region between R.A. 5h. 28m. 9s. and 5h. 32m. 1s. Declination between 4° 36′·8 and 5° 59′·3 South.

Scale—1 millimètre to 30 seconds of arc.

Co-ordinates of the Fiducial stars marked with dots for the epoch A.D. 1900.

Star (.) D.M. No. 1289—Zone – 5°	...	R.A. 5h. 28m. 12·6s.	...	Dec. S. 5° 24′·6	...	Mag. 8·5
„ (··) „ „ 1167 „ 4°	...	„ 5h. 29m. 27·3s.	...	„ 4° 52′·9	...	„ 7·5
„ (∵) „ „ 1334 „ 5°	...	„ 5h. 31m. 20·7s.	...	„ 5° 42′·5	...	„ 7·8

The photograph was taken with the 20-inch reflector on January 15th, 1896, between sidereal time 4h. 12m. and 5h. 42m., with an exposure of the plate during ninety minutes.

REFERENCES.

N.G.C. 1976. G.C. 1179. *h* 360.

Three photographs of this nebula with respective exposures of 15 minutes, 81 minutes, and 205 minutes were published, together with descriptive matter concerning them, in *I.R. Photos.*, pls. 15, 16, and 17. The one No. 3 was taken with the 20-inch reflector on February 4th, 1889, with an exposure of 205m. Another photograph with an exposure of 7h. 35m. was taken between the 3rd and 8th February, 1894, and was published in *Knowledge* on the 1st April, 1897.

The photograph now reproduced is presented for comparison with those referred to above, and also as a practical illustration of the unreliability of any judgment that may be founded upon time intervals only in reference to photographic results. The illustration is only one of many others similar in character which I have met with. For example, the photograph annexed was taken with an exposure of only ninety minutes, yet it shows all the faint stars and nearly all the nebulosity seen on other plates that have been exposed to the same area of the sky during various intervals of time between three hours and seven and a half hours. Sometimes a plate exposed for half an hour, or forty minutes, will compare very favourably with others exposed for four hours.

What is the cause of these differences? Is it due to unequal sensitiveness in the films, or to varying atmospheric conditions, or to temperature, or to instrumental causes? Unequal sensitiveness of the films may be dismissed, for they are always tested for equality in that respect. The atmospheric conditions are probably the chief cause. Unfortunately there is at present no reliable method available by which the photographer while at work can ascertain the probable effect of the atmospheric conditions upon his work.

It may be that, in addition to the many degrees of obscuration in the sky by atmospheric conditions, there exists either within or beyond the confines of the solar system, clouds of meteoric, cometic, or dark nebulous matter that are frequently interposed between us and the objects we are photographing; thus intercepting part of the light on its way to us from the stellar regions. Another consideration that I should lay stress upon is that with exposures of many hours' duration in the 20-inch reflector there is not the increase which we should expect in the number of stars and extent of nebulosity shown upon the plates. As I have already suggested, on page 20 *ante*, we have probably reached the limit of the power of the photographic method in the records we have already obtained, and that consequently there is a void in space between the *Galactic Circle* and the next Celestial System beyond it.

It might be urged that exposures of the most sensitive plates to various parts of the sky during intervals of (let us say) 100 hours on successive clear and suitable nights, would show the existence of a larger number of faint stars, and more faint nebulosity, than has hitherto been proved. If this could be demonstrated it would remove farther still the supposed limit of the celestial space to which I am referring.

The evidence which I now submit in support of the inference that a limit of photographic power has been reached, is to be found upon my plates which have been exposed to the light of special objects in the sky during intervals of eight to twelve hours in the 20-inch reflector. The films on each of the plates so exposed have been darkened by the diffused light of the sky by an amount equal to the density of star images of 17th to 18th magnitude, and it is evident that as soon as the darkening of the films generally is equal to the density of any given star-images it would not be possible to develop the images of any stars or nebulosity that are fainter than that limit.

Time, together with close observations by aid of the most powerful telescopes, and the best possible photographic work, must finally determine the accuracy or otherwise of these inferences, which are drawn from the photographs now published, and from the original negatives.

The inferences and the statements here made cannot be either proved or disproved by mere academical discussions; they must be *demonstrated* by further evidence. If the possessors of instruments can furnish indisputable evidence by eye-observations, or by any form of photographic appliances that there exist fainter stars, or fainter real (not spurious) nebulosity, in the region of the *Pleiades*, comprised within a circle of one degree radius with *Eta* as the centre, than are shown on my photographs taken with the 20-inch reflector and exposures up to twelve hours' duration, such demonstrations would be most desirable in the solution of this very important astronomical question.

Again, and in continuation of the foregoing paragraph, if the means there detailed were applied to the study of the stars and nebulosity in the region of the Great Nebula *M.* 42 *Orionis* that are comprised within a circle having a radius of one degree with any star in the *Trapezium* as the centre, the results would be most desirable in confirmation, or otherwise, of those obtained in the case of the *Pleiades*.

These are test questions that would well repay investigation by those astronomers who have command of the most powerful modern instruments yet constructed. I consider that I have already exhausted the powers of my own instrumental equipment, and have by their aid reached my limit. Can others produce indisputable evidence that there exists in space fainter stars and fainter nebulosity than I have succeeded in proving?

Mere prolonged exposures of plates to the sky will be of no avail in the settlement of the question. What we want to see, in the first instance, is the proof that fainter stars and fainter nebulosity exist in the two regions mentioned in the preceding paragraphs; afterwards further tests can be applied to other parts of the sky.

PLATE XXVI.

Nebula M. 1 Tauri.

R.A. 5h. 28m. 30s. Dec. N. 21° 57′·0.

The photograph covers 15′·3 in R.A. and 21′·8 in Declination, and was taken with the 20-inch reflector on January 25th, 1895, between sidereal time 4h. 30m. and 5h. 30m., with an exposure of the plate during sixty minutes.

Scale—1 millimètre to 8 seconds of arc.

REFERENCES.

N.G.C. 1952. G.C. 1157. *h* 357.

Sir J. Herschel, in the G.C., describes the nebula as very bright; very large; extended in the direction 135° ±; very gradually a little brighter in the middle; resolvable. A drawing of it is given in the *Phil. Trans.*, 1833, pl. XVI., fig. 81.

Lord Rosse (*Obs. of Neb. and Cl.*, p. 47) describes it; and in the *Phil. Trans.*, 1844, pl. XVIII., fig. 81, a drawing is given.

I.R. Photos., pl. XIV.

Lassell, in the *Mem. R.A.S.*, Vol. XXIII., pl. II., fig. 1, gives a drawing of it.

The photograph shows the nebula to be elongated in *s.f.* to *n.p.* direction; irregular in outline, and somewhat resembles an island, with deep bays at intervals round its margin. The original negative shows mottling and rifts in the nebulosity, and that one of the rifts curves near the *n.f.* margin; another extends across from the *n.f.* to the *s.p.* side, with a star of about the 14th magnitude at its centre. There are also some star-like condensations involved in the nebulosity.

NEBULA N.G.C. 2237-9 MONOCEROTIS.

PLATE XXVII.

Nebulous Region round the Cluster N.G.C. Nos. 2237-39 Monocerotis.

The photograph covers the region between R.A. 6h. 24m. 16s. and R.A. 6h. 29m. 13s. Declination between 4° 20′·8 and 5° 56′·5 North.

Scale—1 millimètre to 24 seconds of arc.

Co-ordinates of the Fiducial stars marked with dots for the epoch A.D. 1900.

Star		D.M. No.		Zone		R.A.		Dec.		Mag.
Star	(.)	D.M. No.	1271	—Zone + 5	...	R.A. 6h. 25m. 30·7s.	...	Dec N. 5° 51′·7	...	Mag. 8·7
„	(··)	„ „	1288	„ 4	...	„ 6h. 25m. 44·7s.	...	„ 4° 35′·2	...	„ 8·3
„	(∴)	„ „	1302	„ 5	...	„ 6h. 28m. 49·4s.	...	„ 5° 12′·5	...	„ 8·3

The photograph was taken with the 20-inch reflector on March 5th, 1899, between sidereal time 6h. 28m. and 9h. 13m., with an exposure of the plate during two hours and forty-five minutes.

REFERENCES.

G.C. 1420. N.G.C. 2237-38-39. *Ast. Nach.*, No. 2918, pp. 253-4.

Astron. and *Astro-physics*, Vol. XIII., p. 178.

The photograph shows the nebula in complete form and with much detail available for present examination, and for future comparison, with a view to detect changes and motions amongst the stars and the nebulosity.

The nebula extends about 77 minutes of arc in *south following* to *north preceding* direction, and about 67 minutes from north to south; it is a cloudy mass broken up into wisps, streamers, and curdling masses densely dotted over with stars, which also extend widely over the regions surrounding. There are many dark areas with few, if any, stars in them, and they are also free from nebulosity; other areas have both faint stars and faint nebulosity in them which the photograph delineates to a degree of faintness equal to that of the light of stars of about the 18th magnitude.

Some remarkable black tortuous rifts meander through the nebulosity in the *north preceding* half of the nebula; their margins are sharp and well defined in the midst of dense nebulosity; they are like cleanly cut cañons; they are not due to instrumental or chemical or other defects, for I have ample proof of their reality.

The stars or star-like condensations involved in the nebulosity are easily distinguishable from the apparently fully formed stars that are strewn over the surface of the nebula; the evidence is strong that the stars are between the earth and the nebula, but at present I cannot produce demonstration of the accuracy of this inference.

Plate 28.

CLUSTER M. 13 HERCULIS—*a*.

CLUSTER M. 13 HERCULIS—*b*.

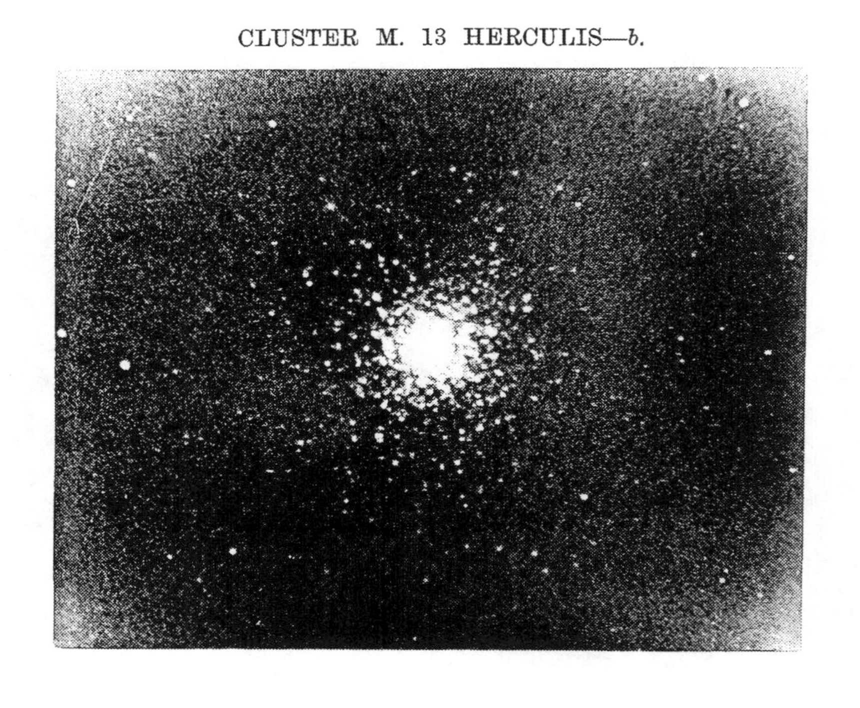

CLUSTER M. 14 OPHIUCHI.

CLUSTER M. 2 AQUARII.

PLATE XXVIII.

Cluster M. 13 Herculis.

R.A. 16h. 38m. 6s. Dec. N. 36° 39′·0.

The photograph covers 20′·4 in R.A. and 16′·3 in Declination, and was taken, (a) with the 20-inch reflector on May 28th, 1895, between sidereal time 14h. 51m. and 15h. 51m., with an exposure of the plate during sixty minutes; (b) was taken with the 20-inch reflector on June 15th, 1895, between sidereal time 16h. 37m. and 16h. 42m., with an exposure of the plate during five minutes.

Scale—1 millimètre to 12 seconds of arc.

REFERENCES.

N.G.C. 6205. G.C. 4230. *h* 1968.

Sir J. Herschel, *Phil. Trans.*, 1833, p. 458, pl. XVI., fig. 86. Lord Rosse, *Phil. Trans.*, 1861, p. 782, pl. XXVIII., fig. 33, and *Obs. of Neb. and Cl.*, p. 150. A. C. Ranyard, *Knowledge*, 1st May, 1893, pp. 90-93.

I.R. Photos., pl. XXXIV., p. 98.

The photograph (b) that was taken with an exposure of five minutes shows the stars in the central part of the cluster to be involved either in faint nebulosity or in atmospheric glare caused by the light of the stars, and some of them appear to be deformed in outline because of the overlapping of two or more star images. On the other photograph (a), which was exposed during sixty minutes, the nebulous obscuration extends farther from the centre.

The general configuration of the stars are suggestive of their development from a spiral nebula, and if we refer to the series of photographs of spiral nebulæ on the Plates numbered XI. to XVIII. *ante*, we shall see upon them striking confirmation of the probable accuracy of this inference; the forms of the convolutions still remain visible in the arrangement of the stars, whilst the nebulous matter appears to have been absorbed, and the nuclear condensations at their centres account for the dense aggregations of stars.

We can easily realise that the central nebulosity in the great nebula in *Andromeda*, and in some of the other spiral nebulæ, would furnish ample material for the formation of such clusters as M. 33 *Herculis*, M. 14 *Ophiuchi*, and M. 2 *Aquarii*.

PLATE XXVIII.

Cluster M. 14 Ophiuchi.

R.A. 17h. 32m. 21s. Dec. S. 3° 11′·2.

The photograph covers 20′·4 in R.A. and 16·3 in Declination, and was taken with the 20-inch reflector on August 2nd, 1897, between sidereal time 18h. 42m. and 21h. 0m., with an exposure of the plate during two hours and eighteen minutes.

Scale—1 millimètre to 12 seconds of arc.

REFERENCES.

N.G.C. 6402. G.C. 4315. *h* 1983 = 3698.

Sir J. Herschel, in the G.C., describes it as a remarkable object; a globular cluster; bright; very large; round; extremely rich; very gradually much brighter in the middle, resolvable into stars 15 to 16 mag.

Lord Rosse (*Obs. of Neb. and Cl.*, p. 149) states that the stars are very close together, and most of them very small.

The photograph shows the cluster, with curves and lines of faint stars radiating in all directions outwards from the dense centre; and that in their general aspect they resemble a spiral nebula without any nebulosity, excepting at the centre. The negative shows curves, lines of stars, as well as vacancies within the nebulous centre.

The black crooked line on the *south preceding* side, and the faint line pointing due north, near the cluster, are due to defects in the film.

PLATE XXVIII.

Cluster M. 2 Aquarii.

R.A. 21h. 28m. 19s. Dec. S. 1° 15′·9.

The photograph covers 20′·4 in R.A. and 16′·3 in Declination, and was taken with the 20-inch reflector on October 30th, 1891, between sidereal time 21h. 57m. and 23h. 25m., with an exposure of the plate during eighty-eight minutes.

Scale—1 millimètre to 12 seconds of arc.

REFERENCES.

N.G.C. 7089. G.C. 4678. *h* 2125.

Sir J. Herschel, in the G.C., describes it as partly resolvable; a globular cluster; bright; very large; gradually, pretty much brighter in the middle; resolvable into stars; stars extremely faint. A drawing of it is given in the *Phil. Trans.*, 1833, pl. XVI., fig. 88, and 1844, pl. XVIII., fig. 88.

Lord Rosse (*Obs. of Neb. and Cl.*, p. 162) describes the cluster, and gives measurements of the position angles, and distances of some of the stars.

The photograph shows the cluster with a large central mass of nebulosity so dense as to obliterate the star-images; but the faint stars surrounding it are arranged in a manner suggestive of their origin from a spiral nebula.

The three clusters depicted on Plate XXVIII., and there are others of a similar character but not yet published, are strongly suggestive of, if they do not indisputably prove, that the same principle of aggregation has been in operation to cause the origin and development of each of these clusters, and I have not been able to detect any clearer evidence of their origin than that of development from spiral nebulæ.

If the conclusions are correct which I have drawn from the photographic evidence submitted in the foregoing plates and descriptive matter, the following would probably be the order of stellar evolution, commencing only at the epoch of *reconstruction* and having no reference to the origin of matter itself.

(1) Dark or light aggregations of matter in globular, cometic, meteoritic, or dust-like form, and in gaseous clouds scattered about, isolated, in space.

(2) Collisions between any two or more of such bodies or aggregations.

(3) Re-combination of the materials after collision into nebulæ, mostly of the spiral type, and then into stars.

(4) Arrival again at the epoch of maturity.

(5) Decay, and then return to the epoch of quiescence preparatory to undergoing another cycle of wreckage and of re-constitution.

These cycles are within the range of our comprehension, since we may see for ourselves in the earth by the study of the materials which form its crust.

There can be no suspicion of errors of judgment on the part of competent geologists in proving that processes of breaking up of the old materials consisting of organic and inorganic compounds by the continuous application of diverse known forces, and reconstructing from them new forms of earth-material and of life developments, are, and have been, in progress ever since the geologist's *Cambrian* or *Laurentian* epochs. May not, therefore, the infinitely larger, and older, processes of breaking up and re-combining the materials of Celestial Systems be the same in principle as that which we can see taking place on the earth on a relatively microscopic scale?

What we need is intellectual development to enable us to view these physical developments in the proportions of their respective magnitudes and time-intervals.

The breaking up, by collision, of two solar systems, if the process could be viewed, or photographed, from the distance of *Sirius* would be an event insignificant to sight and probably too feeble in its light intensity to imprint more than a small splash on a photographic plate.

Printed in the United States
By Bookmasters